COLD BEANS OUT OF A CAN

COLD BEANS OUT OF A CAN

From Teenage Aircraft Mechanic and Pilot to Apollo Engineer

HARVEY A. SMITH

SMALL BATCH BOOKS
Amherst, Massachusetts

Printed in the United States of America.

Library of Congress Control Number: 2019911529

ISBN: 978-1-937650-99-5

Designed by Lisa Vega

493 South Pleasant Street
Amherst, Massachusetts 01002
413.230.3943
smallbatchbooks.com

For Cecile Simone Morin Smith,

Lori, Edward, Peter, and Dwight

CONTENTS

Acknowledgments

I would like to thank Allan Edmands for his many valuable suggestions, Allison Gillis for her eagle-eyed editing, Lisa Vega for her excellent design work, and Fred Levine for his always excellent, patient, and knowledgeable guidance. Thanks to everyone at Small Batch Books for transforming the words penned by an engineer into a finished and polished book!

I would also like to thank Scott Sacknoff, publisher of *Quest: The History of Spaceflight*, for his help in recovering three lost images.

Astronaut Russell Schweickart using the Hamilton Standard backpack for the first time in space. (NASA photo)

PROLOGUE

Keeping a Man Alive in Space!

The year was 1969. Astronaut Russell L. "Rusty" Schweickart emerged from the lunar module (LM) and put his feet in the "golden slippers," the restraints to hold himself in place. Schweickart was using the *Apollo 9* backpack for the first time in space.

Schweickart and the backpack were wired up with **telemetry**,[1] which was received in an office at Hamilton Standard, where I worked as a design engineer and manager. Schweickart was also shown on a TV about two hundred feet away. This backpack had been tested and analyzed in every conceivable way by Hamilton's Space and Life Department for years, but this was the first time it was actually being used in space.

We raced down to the telemetry room and stuck our heads in. My heart was racing as the operator turned around and announced, "Everything is fine."

I was so wound up, I did not know what else to do, so I went back to my desk and shuffled some unimportant papers to try to wind down. There were no high fives as seen on TV. We just knew that everyone at Hamilton had done their job well. The backpack was working perfectly; it was keeping a man alive in space!

1. Boldface indicates that the term is defined in the glossary.

At this time, I was leading a group of fifty engineers, working on all the backpack changes for the final version, which was to go to the moon. So how did I get to be a manager with fifty engineers on NASA's Apollo program? This is my story.

CHAPTER ONE

Working "in the Hangar" and Flying at Fifteen

In the summer of 1951, Bert Marona, manager of the airport at West Brookfield, Massachusetts, hired me to mow the grass on the airplane parking area. I was fifteen years old. He had had several boys who had not worked out, and I decided to ask him for the job. At first, he was not interested, but I persisted. I still remember Bert sitting there in the airport office with a lot on his mind. I asked him again about the work, and he finally answered, "Okay," almost as an aside.

We lived in Brookfield on a small farm. We raised most of our own food. My mother did a lot of baking and canning. My father worked at the American Optical Company in Southbridge as a laboratory technician. He was very supportive of the Boy Scouts and camping, and he donated the use of some of our twenty-five acres for a permanent campsite, where my friends and I had some great times. We were all raised in what today is called "free-range parenting"; we were free to go off camping unsupervised and to drive our old cars around our parents' properties. Things were different in those days, with more open land and fewer restrictions.

I had spent some summers at my grandparents' farm in Wilbraham, Massachusetts, where my grandmother told me, "Harvey, if you work hard, you will always have a good job and do good." Consequently, I mowed the airport grass like a demon, using an ancient mower; I clearly remember that it had a patent date of 1918 on it, and it had a single steel tiller-style handle. I did not stop for water in the hot summer days; I stopped only to refuel.

As I worked, I saw Bert looking out of the hangar at me from time to time, but I didn't think much of it. After about three weeks, he came over to me and said, "How would you like to work in the hangar?"

"Doing what?" I asked.

"Oh, cleaning parts and tools, removing plates, and so on." He then added, "We are not going to let you hurt anything." Then: "I will give you a little money [and it was little], but you will learn to fly and get your license."

I do not remember exactly what I said, but it amounted to "I am in!" I knew this was a big opportunity, and I wanted to make the most of it!

I already knew the airport routine pretty well. When I was twelve, I had started hanging around there, getting rides and helping to push planes around and tie them down. I had two uncles who had been World War II fighter pilots and an aunt who had been a WASP (Women Air-force Service Pilot), flying all kinds of aircraft on ferry service and towing targets. It had seemed natural for me to show up at the airport.

Uncle Luther had joined the army air corps (precursor to the U.S. Air Force) before the war. During the war, he became a fighter pilot in Europe. He rose to squadron commander and lieutenant colonel. While strafing the enemy with his P-51 Mustang, he was shot down, was captured, and spent a year in a German POW camp. After the war, he was assigned to work on the rearmament of Germany. He personally brought the German defense minister over to the U.S. to visit various aircraft plants and evaluate different aircraft.

Uncle William had been an army private at Pearl Harbor when it was bombed, an event that he said, in his understated way, was "awful." Shortly after the bombing, he responded to a posting that asked for anyone with flying experience who wanted to go to pilot training to sign up. The posting did not specify how much flying experience was required. William had had five hours in a Piper Cub in Hawaii, which he had paid for out of his private's pay. Apparently, that was enough, because he began his pilot training a few days later. He started out flying P-40 Warhawks (he later told me they were "dogs"). At some point, he switched to P-47 Thunderbolts, which he flew until the end of the war. He shot down two and a half airplanes (another pilot was credited for firing rounds into the third plane) and was shot down himself twice, but each time he bellied in instead of bailing out. He flew missions all the way from Africa, up the Italian peninsula, and into Germany, rising to the rank of captain and squadron commander. After the war, he told me, "We could not wait to get in that plane in the morning and go bag 'em." According to a family source, he flew 225 missions and could have come home years earlier but chose not to.

Aunt Henrietta had taken some flying lessons at the Springfield (Massachusetts) Airport on her salary as a secretary and decided she wanted to do something for the war effort. She applied to the WASP program and was accepted. After completing her training, she flew ferry duty for single-engine aircraft, twin-engine bombers, and even a four-engine bomber. She was also assigned to tow-target duty: She towed targets on a two-thousand-foot cable for ground artillery to shoot at. This duty was very dangerous; even with such a long cable, a tow airplane was occasionally hit!

Another uncle earned a private pilot's license at the age of eighteen before going into the service. Uncle Alvin soloed in only four hours, which was an unheard-of achievement; at that time, the usual minimum instruction time to solo was eight hours of dual instruction. Upon enlisting in the air force, Alvin was a radar man on F-82 Twin Mustang aircraft.

These people were—and still are—my heroes!

My dad paid for my first airplane ride in a Cessna 140 flown by the airport owner, Gus Willet, for what was probably my twelfth birthday. I remember that the height did not bother me, as it does some people, and I just looked at the scenery and watched what Gus was doing to operate the airplane.

My second ride, when I was also twelve or so, was by a gentleman I knew only as Ski in a 40 hp Taylorcraft, which was a high-wing aircraft that was very economical to operate due to its low fuel consumption. It seated two people side by side comfortably. In the center of the instrument panel was a big, round instrument approximately ten inches in diameter; when Ski opened the throttle, a huge needle went around, giving me the impression that something spectacular was happening. The instrument had some smaller instruments within it, probably to indicate oil pressure and oil temperature. Unfortunately, he was killed shortly thereafter when some other aviators came over his property and struck some high-tension wires. He ran out, tried to free them, and was killed, while they were unharmed.

One of the airport regulars, Gordon Richards, had a Luscombe, a two-seater 65 hp plane, the first small aircraft whose fuselage was entirely made of sheet metal and whose wings were fabric covered in the days when most light aircraft had a welded-tube, fabric-covered construction. Gordon's favorite gag was to start a slight roll while saying, "Did you ever do a snap roll?" (A **snap roll** is one rapid 360-degree revolution about the longitudinal flight axis.) By the time he finished his question, you were already all the way around, 360 degrees.

On the first morning of my new job in the hangar, I rode my bike four miles to the airport, arriving one hour early, at 7:00 a.m. I opened the hangar doors, swept out the whole hangar, and wiped down all the mechanic tools. I was just finishing the tools when Bert stepped into the hangar. His head snapped back, and he stopped dead in his tracks and looked around at what I had done. He obviously was very pleased.

The first thing Bert did was to teach me that *everything* on an airplane, except for the upholstery, is *critical*. I was quickly taught to remove **inspection plates**, and I did a lot of crawling down into the belly of an airplane to vacuum it out or to hold a wrench for someone working on the outside. I quickly understood this was serious business. The mechanics were often doing airworthiness inspections, where all external plates are removed and the entire structure and all moving parts are inspected for any signs of deterioration. Criticality was pounded into my head, so when I later got into spacecraft equipment design, it was second nature.

I was taught to count my tools when I went into an airplane and to count them again when I came out because any loose tools could jam the controls. I later found tools left in an airplane by others who had not been trained as well as I had been.

I was also taught where to push on an airplane and where not to. Soon I was removing the **cowlings** that fit outside the engine and that needed to be removed for access. I would then degrease the engine if it needed it. I even did a little spray-painting, and I helped with riveting by holding the **bucking bar**, the solid piece of steel that flattens a rivet being driven by the rivet gun from the other side.

I listened intently to all instructions, and I learned all kinds of things: things such as filing out the nicks in a metal propeller because they were "stress nodes," as they are called. I would later learn in engineering that these nodes are termed **stress concentration factors**, which determine how much stress is increased from localized changes of shape in the base material. I learned that such factors are indeed very important, whether in airplanes or in spacecraft.

Here I was thriving. I always had in mind that my uncles and aunt would be pleased with what I was doing. To me, this was not merely a workaday summer job. This was exciting, and I could not wait to get to work. One day, riding my bike on a downhill section of road, I actually passed an elderly man in a big Chrysler 300. He looked startled and jerked the wheel a little to see me speed by him on my way to work at the airport!

It turned out that Bert Marona was a master mechanic. He was an FAA-licensed mechanic with **airframe** and powerplant (A&P) ratings as well as an FAA inspector. He was also a well-qualified pilot. He had flown innumerable different models of airplanes. Since being in the maintenance business, he often had to go pick up a model he had never flown before. He was always proud of the fact that he had "never put a scratch on an airplane."

FLYING AT FIFTEEN

I started getting flight instruction in Piper Cubs from a pilot by the name of Paul Cormier. The Piper Cub was the Model A Ford of aviation, in which thousands of people learned to fly. It is a very durable, high-wing, 65 hp tandem trainer that is ideal for initial flight instruction at a low cost.

Paul had been an **aerobatics** instructor in the navy for three years, flying Stearman biplanes, one of which he owned on the field. The Stearman is extremely rugged, with a 220 hp engine designed for basic training of military cadets. During World War II, thousands of pilots received basic flight training in Stearmans, and today they are prized for use in air shows and as collector's items. Paul wrote in my logbook "Eager to learn." I was doing exactly what I wanted. What could be better?

Paul had a French Canadian accent, and after a while he told me, "Arve, zee airplane is like zee woman. If you bat her around, she is not good to you. You have to be smooth to her." Paul was a fantastic pilot. When flying his own Stearman, his signature maneuver at the end of the day was to fly down the field downwind, pull up vertically into a 180-degree **wingover** at the far end of the runway, and come in and land. This was no ordinary landing. This was so smooth, we would all watch for the wheels to start turning on the grass. There was no bouncing or even visible deflection of the **shock struts**. All those watching said nothing, but we always knew we were watching a masterful performance!

The runway on the east end of the West Brookfield Airport started on top of a one-hundred-foot hill. So a normal takeoff was down the hill, onto grass, onto a flat, and off. I was trained to routinely land on the first three hundred feet of the runway. I was also trained to recover from spins and engine quits. I have to chuckle when I hear people talk about what a harrowing flight they had on a ten-to-twelve-passenger twin-engine "puddle jumper."

One day near the end of August, Bert said to me, "Harvey, won't you be going back to school soon?"

This struck me like a bolt out of the blue. "Oh no," I told him, "I am going to stay here and work."

However, unbeknownst to me, Bert called my dad and said, "I think I could make arrangements for Harvey to go the aircraft mechanics school where I went in Putnam, Connecticut. They need students."

Next, he arranged an interview for me with H. H. Ellis, the principal of Putnam Technical School, which had formed an Aircraft and Aircraft Engine Maintenance Department approved by both the state of Connecticut and the FAA. Mr. Ellis explained that I would have to live in Connecticut at least on weekdays, a requirement to which I readily agreed.

Bert had definitely taken me under his wing, for which I am eternally grateful! He was my mentor for the next six years. I will always be indebted to him as well as to several other people at the airport. One of the customers who knew him well told me quietly one day that Bert and his wife had lost a baby boy who would now be about my age, and the customer thought that was part of why Bert had taken me under his wing. I am sure the other part was that I had been the last kid who showed up at the airport and worked hard. Bert told me that himself!

I did stop at Brookfield High School and told Principal Henry Card that I was leaving and where I was going. I had a good relationship with Mr. Card. He was also a pilot; he had taken me on a ride a winter or two before in a J3 Piper Cub on skis. We took off and went to the Quabbin Reservoir, where he saw some eagles feeding on a deer carcass and landed alongside. Of course, the eagles took off, and I clearly remember how they left about a six-foot disturbance in the snow with their wing tips.

CHAPTER TWO

Cold Beans Out of a Can

So now at age fifteen, I was enrolled as a student in the Aircraft and Aircraft Engine Maintenance Department at Putnam Technical School, in Putnam, Connecticut. I enjoyed every minute of what I was doing in this three-year program and could see where I was going.

For the first two weeks, I was taken in each day by the chief instructor, William Simmons, and his wife, Anne. After that, Mr. Ellis called me in and told me that he knew a lady from church by the name of Mrs. Sabin, who rented rooms right near the school. I went to see her and learned that she was a widow who had lost her husband in a blasting accident. She was very prim and proper. My room was on the second floor. And it was just a room. There was a bathroom I needed to share with two other roomers. No cooking facilities. Thus, I was to eat cold beans out of a can for three years. Breakfast was three doughnuts at the diner and a cup of coffee.

As regards discipline, Mr. Ellis was from the old school. Each morning he would walk up to the entrance of the school, wearing his three-piece suit and using a cane. Out of respect, everyone stood aside on the walk as he went by. We could do with some of this respect today!

Although the school was not on an airport, the Aircraft Department had a regular aircraft hangar and engine test cell equipped to test the large aircraft engines of the day. When I got there, they actually had in the hangar a whole Curtiss P-36 Hawk fighter, which was the forerunner of the P-40. The P-36 had a Pratt & Whitney 900 hp radial engine (which was replaced with a 1,360 hp Allison in-line liquid-cooled engine in the P-40). The school also had a centrifugal jet engine. There was no formal instruction on either of these pieces, but I did get to study the construction of each in detail, sit in the cockpit of the P-36, and do some "hangar flying."

My goal was to get a diploma in aircraft and aircraft engine maintenance, which would qualify me to take the FAA **airframe and powerplant (A&P) mechanics license** tests. I would also earn a secondary school diploma. I could work on my private pilot's license at West Brookfield Airport whenever time permitted.

For three years, I had no social life at school whatsoever, although I did see friends at home on the weekend. During the first year, my usual transportation back and forth to Brookfield was thumbing, which worked perfectly well in those days.

It never occurred to me to ask my parents to pay for my room or board. My parents were good people, but they certainly were not rich. My mother helped by doing my laundry each weekend. It was always ready, neat and folded, on Sunday. Today most young people expect their parents to provide everything for school. I did not!

WORKING AS A PRINTER'S DEVIL

Next, I needed employment. For several nights after school, I walked all over the town of Putnam, looking for work. Finally, I went into a combination weekly newspaper and printing business called the *Windham County Observer*. I wound up talking to the owner, Stanley Evans, who told me he thought they could use someone after school to do odd jobs. Years later, I would realize that this was another case of someone taking me under his wing and helping me. I learned later that Stanley had been a World War II P-47 aircraft crew chief and could relate to what I was trying to do. I am convinced that there are good people everywhere.

My job title was printer's devil. My duties were a variety of things: Melt the linotype lead in a small furnace, grease the printing press used for printing the weekly paper, and operate a small hand press. I also carried the weekly newspapers to the Putnam post office, which was nearby. The only problem was that Stanley had taken over the business from his father, who was still working on the premises and was right out of the 1880s. He wore heavy black work shoes and a tattered three-piece suit, and he reeked of tobacco. He would announce himself by walking up in back of me and kicking my shoe heel. I wondered if this was also from the 1880s. He typically questioned me on what I was doing, which was following the instructions that his son, Stanley, had given me. He would say something to the effect that Stanley did not know what he was doing and that I should do X, Y, and Z instead. So I would be off doing X, Y, and Z when Stanley would come up and ask me why the hell I was not following his instructions. Then the two of them would meet among the presses and have a shouting match, after which I would go back to following Stanley's instructions.

WEEKENDS BACK AT WEST BROOKFIELD

Saturdays and Sundays, I worked back at the airport for Bert. Sometimes, someone at the airport would fly me back to school after my mechanics work was done, landing at South Woodstock (Connecticut) Airport on Sunday night. I would then thumb into Putnam. Mrs. Sabin was always sitting up, waiting for me on Sunday nights, as she took an interest in how I was doing and she liked to talk. She was always dressed up like the lady she was, sitting in her front room waiting. One evening, she suggested that if I would fill her kerosene tank once a week, she would let me eat dinner with her and one elderly female permanent boarder who ate with her one evening a week. I was really looking forward to a sit-down dinner, but when it actually happened, the two women did not eat enough to keep a small bird alive. As I recall, it was a little bit of tuna fish on lettuce, a few crackers, and tea. So I ate what was offered, went upstairs, and ate my usual cold beans in addition.

I generally had no problems thumbing, but I did have a couple of memorable episodes. One Friday night, I was just leaving Putnam for Brookfield when I got a ride with a man in on old car. As soon as I got in, I knew something was not right. He just looked straight ahead and said nothing. He then started to accelerate into a curve ahead, drifting into the opposite lane, and barely following the grassy edge of the wrong side of the road. He was "drunk as a skunk," as they say. Just then a Connecticut state trooper was in back of us with the siren on. The driver seemed oblivious to the siren, and I thought, Oh my God! I am going to be in a police chase. I yelled at the driver, "The police are behind us," but he did not react. I kept yelling, and finally, he pulled over.

The trooper stepped up to his window, which was open, reached in, and pulled out the key. He said to both of us, “Let’s go,” motioning to the cruiser.

“I was just thumbing,” I quickly said in a high-pitched voice.

The trooper said, “Take off,” which I promptly did.

WHAT I LEARNED IN SHOP AND IN MY CLASSES

At school, I threw myself into the work and did my very best. I was enjoying it! The school was organized into alternate weeks of shop and classes. The week of shop covered engine overhaul, welding, sheet metal, and all the other aspects of aircraft maintenance. This was a first-class school, where again, I was taught that every aircraft part is *critical*! During the first six weeks or so, we all had some machine shop training so we would at least become familiar with what happens in a machine shop. We were shown what job each machine did, and then we were assigned to make some simple parts, such as metal clamps that we could keep in our toolboxes.

The week of classes included drafting, math, and physics. All the instructors were superb! The drafting instructor, Mr. Morehouse, was a graduate aeronautical engineer. I had always wanted to build an ice-boat propelled by a small motor of 20 hp or so, and I made up some sketches and brought them to Mr. Morehouse. I asked him if he could calculate the speed it could reach. He did so, explaining the various steps he went through to get the answer. He explained that he had to find a coefficient for drag for the iceboat as a whole aerodynamically and then, using standard equations, compare that coefficient against the thrust produced by a 20 hp engine and propeller. The friction of the ice would be negligible. He calculated that it could reach seventy miles per hour! I was mightily impressed, and I think that was part of why I later studied aeronautical engineering.

Mr. Huss was a college graduate who had worked for Pan Am Airways as an aircraft mechanic. He was a very good communicator. He instructed us in all phases of aircraft and engine overhaul, including welding, sheet metal, painting, and electrical circuitry.

From books and lectures, we learned the fundamentals of the four-stroke engine. We learned the basic four-stroke cycle and the valve-ignition timing concept that engines run on: The first stroke of the piston downward is with the intake valve open, drawing in the air-fuel mixture. The second stroke is upward with both valves closed, compressing the mixture. At the end of this stroke, ignition occurs, and the controlled explosion drives the piston down on the third stroke with both valves closed. On the fourth and upward stroke, the exhaust valve is open, and the piston clears out the exhaust products.

During Engine Overhaul class, each student would do one step at a time—for example, fasten two nuts and torque them to factory-specified values—and an instructor would check what the student had done and sign off on it. When the engine was all overhauled, we would move it into a test cell and run it a prescribed period of time. We were also taught to do a **magnaflux**, or **magnetic particle inspection**, to test all steel parts to look for cracks and to inspect all parts dimensionally to factory specifications with mechanical measuring instruments. We did all the work on live airplanes and engines that would be returned to service. The only difference was that an engine overhaul would take two years at the school because of the strict oversight and the need for step-by-step signoffs, compared with only two weeks in a regular aircraft maintenance shop.

One customer was a crop duster operator who brought in an engine every year for major overhaul. In fact, one morning when I was thumbing to school, the driver of a small truck stopped to give me a ride and asked me where I was going.

"I am going to Putman Technical School," I proudly said.

"So am I," he said. "I am bringing in an engine for overhaul!"

It is sad that a great many parents today have a problem sending their children to a technical school. It is a shame that even today, there is a stigma attached to technical or trade schools. The graduate of one of these schools can actually *do* something, such as build a house.

SOLOING DURING SUMMER BREAK

In the summer of 1952, Bert decided to get me soloed. He told his partner, Ray Parker, to work with me and get it done. The Aeronca Champ was similar to the famous Piper Cub in that it was a high-wing 65 hp two-seater trainer. The Champ could be flown solo from the front seat, however, whereas the Piper Cub had to be flown from the back seat due to weight and balance considerations.

I guess Ray felt he had to get me mad for some reason; he had me fly around the pattern a few times, criticizing every little thing I did. When we pulled up to the hangar, he got out and demanded, "Do you think you can get this around the field without smashing it up?"

Of course, I said yes, and then I soloed. The date was August 26, 1952, and I was sixteen years old. When I went home that night, I ran into the living room and said, "Dad, I soloed."

My father did not exactly share my excitement, but he did not say anything. Now that I am the father of four children myself, I know what he was thinking. All parents are concerned about the safety of their children, and the general public is taught that aviation is dangerous. Later, when I got my license, he went for a ride with me.

The Aeronca 7AC Champion, the same model I soloed at age sixteen. (Courtesy of Ken Stoltzfus)

CHAPTER
THREE

Putnam Tech and Cross-Countrying

BARNEY AND THE .45

For all the Saturdays and Sundays of the next two years, as well as during the summer of 1953, I worked back at West Brookfield Airport for Bert. But when I returned to Putnam after my soloing summer of 1952, I found work at a gas station–garage in North Grosvenordale, Connecticut, owned by a man named Barney Girard. At the very first, Barney looked like he could be a rough individual. During my first fifteen minutes on the job, he was showing me how to run the cash register. He had decided to call me Smitty.

"Smitty," he said, "if you get robbed, give them all the money. Do not argue." Then he pulled open a drawer at chest height, took out a .45 caliber handgun, and said, "See this? It is ready to go. Shoot them right in the back." Those were my instructions, and he was *not* kidding!

Actually, it was not a high crime area, and I never got robbed. Barney also took me under his wing, but I never have reconciled how he could instruct a sixteen-year-old to shoot someone in the back.

After a few weeks, Barney gained confidence in my ability to "run the front," meaning to manage the gas. So one day when I came to work, he said, "Smitty, I am going downstairs to have a little nap. If my wife calls, tell her I am gone with the wrecker."

I needed a part and had to go downstairs. Barney had a nice little sleep area on top of some oil cases, with thick blankets on them, next to the furnace. He was sleeping like a baby! This got to be the regular routine as long as I worked there.

APPLYING OUR TECHNICAL LEARNING

Here is a particular memorable task I recall from school: Another student, Arthur Boucher, and I were assigned to "rig" a Stearman PT-17 biplane. This means we were to set all the angles of the wings properly by using levels and adjusting wires. We were given a Boeing manual with all the factory specifications, and we started in. First, we installed what is called the center section, located at the center of the upper wing and above the front cockpit. Once we had that in place, we installed the lower wings by hanging them on support wires. We then adjusted the support wires to factory-specified values for **wing angle**, as viewed from the front and the end of the wing. Then we installed the upper wings, using the vertical struts to support them on the lower wings.

According to our measurements, one **wing strut** was out of adjustment by a massive amount, but we were reluctant to bring it up. As inexperienced students, we worried that we might be wrong. Finally, we did bring up the misadjustment to our instructors, and they agreed it was off. We corrected the problem, which would make the plane fly much better.

Later, we disassembled the airplane and towed it to South Woodstock Airport, where we reassembled it, so the owner could fly it. He said it flew really nice!

A Stearman PT-17 rigged by Arthur Boucher and the author. Arthur is standing, second from the left; the author is kneeling at the left. The others were needed to lift the wings in place and hold them on assembly day.

One day, we were given books on aircraft engines and told to read them. I had a little time before I had to go to Barney's, so I stayed in my room, studying one of the books. I was reading very intently about fundamental four-stroke-engine theory and engine design, including valve and ignition timing, when I looked up and saw that it was dark. *Dark!!* I had become so engrossed in the book that I had forgotten to go to work!

All the next day, I thought, What will I tell Barney? How do I explain not showing up? I decided to just tell him the straightforward truth, which I did.

"No problem," he said. "I believe you." In his own way, he was telling me that he knew I was very intense, and he believed I could get wrapped up in a book. His faith in me made me feel good.

The author at Putnam Tech with a 65 hp Continental engine.

DESIGN SIMPLICITY (AND A YOUTHFUL MISADVENTURE)

During my second year at Putnam Tech, I was able to scrape together enough money to buy a Model A Ford Coupe. Its marvelous design simplicity would influence me as a design engineer later on. The best and most reliable designs, whether with cars or with spacecraft, are the simplest, with the fewest parts. For example, the Model A had no fuel pump; instead, fuel was fed to the engine by gravity. There were no ignition wires; instead, stamped copper strips worked so well, you could have water all over the engine and it would still run. The choke was a straight metal rod from the dash to the carburetor. It still works fine ninety years later.

So next comes a youthful misadventure with a Model A Ford that turned into a design education lesson for me! During Christmas vacation in my second year, I got together with two lifelong friends and we decided to take my Model A out on the ice at Brookfield's North Pond. When we got there, we saw that the ice was safe only along the edge and out about three hundred feet, so we decided to drive along the edge for a while. I sat in the rumble seat with Joel Bolshaw while Richard Mundell drove. Rich did some spin-arounds, and we had a great old time. But that soon got boring, so we decided to go up the road, come down the beach with a good head start, and do some major spinning around. Unfortunately, we spun out onto the unsafe ice and went through, into about five feet of water. We left the car there temporarily.

Simplicity, the Model A Ford engine. (Courtesy of Volo Classic Cars)

Since Rich had been the driver and I was already back at school, a very unhappy Massachusetts State Police officer visited him and his father. Consequently, we came back the next weekend with Rich's father, bringing along a telephone pole and **chain falls**, and we managed to get the car out of the pond and up to Joel's house, which was nearby. A few days later, I drained the oil and water out, put in fresh oil, and drove the car home at top speed.

There is not a car made today that you could do this to, due to all the electronic controls and its general complexity. Simply stated: This is the way to design things!

According to one reasonable estimate, there are at least a half million Model A Fords still in existence, ninety years after they were built. The engines were so reliable that they were used in homebuilt airplanes in the 1930s, some of which are currently licensed with the FAA and are still flying.

CROSS-COUNTRIES

On September 4, 1953, I flew my first **dual cross-country** from West Brookfield Airport to Norwood, Massachusetts. I was seventeen years old. In a dual cross-country, you go on a preplanned cross-country flight with an FAA-certified instructor to a destination sufficiently distant to meet the distance requirement. In my case, the requirement was one hundred miles minimum. I went with Gus Willet, the airport owner and former World War II pilot. In 1953, of course, there was no electronic navigation equipment for small airplanes. You simply drew a straight line on the map where you were going and started looking for landmarks along the way.

It turns out that Gus had a habit of falling asleep in the back, and with me flying that day, he was asleep back there. I was watching the landmarks intently, and everything matched the map—until I got to where the map showed a power line crossing a lake. I found the lake, but no power line, and I overreacted. I turned around and yelled, "Hey, Gus! We are lost!"

Gus had been sleeping with his head against the Plexiglas side window. When he jolted awake, all he saw was water! He came to, sat up straight, and looked all around twice in an instant. The he breathed out heavily and said disgustedly, "Oh, Christ! There is the GM plant right over there."

Since we were headed east and we were near the Atlantic Ocean, I am sure what he must have thought for a few seconds was that this raw student had chugged out past the shoreline and out to sea—and that we would run out of gas soon and plunge into the drink!

During one of my weekends at West Brookfield Airport, just prior to my graduation from Putnam Tech, I completed my solo cross-country, to Marstons Mills on Cape Cod. It was just one hundred miles, which was the required distance. I got there okay, got my logbook signed, and started back. By now it was midafternoon. A strong headwind breeze had come up, so much so that I was not making much progress; in fact, the cars on Route 6 below were passing me. For a few minutes, I started thinking about a backup plan—going back to Marstons Mills for a while until the wind quieted down—but I continued inland, and the wind quieted down for the rest of the way.

MORE APPLICATIONS OF OUR LEARNING

During my time at Putnam Tech, I was fortunate to meet Admiral Luis de Florez, the gentleman who had won the Collier Trophy, aviation's highest award, for the development of aviator training devices during World War II. His ground training simulators, or "synthetic" training setups, simulated every in-flight emergency on the ground rather than in the air.

He brought his personal airplane into Willimantic, Connecticut, and we Putnam students went down and worked on it. He owned the South Woodstock Airport, and on occasion, he would arrive, flying a World War II navy fighter. In talking to him briefly, I saw that he was very down to earth. I learned years later that he had sixty patents in aviation safety and fuels and was an MIT grad; one of his patents was for the cracking and distillation of hydrocarbon oil. He wrote a letter to the school thanking the students for their fine work on his airplane. It was signed simply: Luis de Florez, engineer!

I also learned welding at school, and I loved doing it. I mentioned this to Barney, and the next thing I knew, there would be several welding jobs waiting for me when I came to work. This was okay with me. Barney would "run the front" because my welding work was more profitable than gas sales.

As graduation time approached, however, Barney's business inexplicably fell off. Though he did not need me anymore, he kept me on for the last two months because he knew I needed the job to finish school. I had worked for Barney for two years, and he turned out to have a heart of gold.

GRADUATION

My graduation was coming up! I got on a bus and traveled to the Anderson-Little clothing outlet in Rhode Island and bought myself a new suit.

On June 30, 1954, I graduated from Putnam Technical School. I received my diploma from the hand of H. H. Ellis himself, and I have never been prouder! For Mr. Ellis, the school was his life. After the speakers had finished carrying on, it was time to award the diplomas; he smiled and said, "Now for the pièce de résistance."[2]

My dad, Bert, and several people from West Brookfield Airport showed up at the ceremony. When it was all over, my dad said, "You have to come home with us tonight." This was a surprise to me, as I had planned to finish up things in Putnam the next day. He repeated, "You have to."

When we got back to the airport, I saw that all the airport people, pilots and aircraft owners, had chipped in and bought me a new roll-around mechanic's tool chest that I would need for my work. My grandfather was there, all smiles. Not much was said, but I could see and feel the pride in everyone present. I was their education project, and it had worked! Someone said, "Speech!" but I could not get out a single word.

These days, I often return the favor of offering encouragement and guidance to young people working on their education, just as these men did for me. I enjoy helping.

---

2. Unfortunately, Putnam Technical School suffered major damage in the 1955 flood, which washed away the hangar and half of the school. Ironically, the higher-ups had always wanted an aircraft school on an actual airport, so the new H. H. Ellis Technical High School and airport was built in nearby Danielson, Connecticut. Years later, the State of Connecticut decided to relocate the school to a more geographically central location, Brainard Field in Hartford, where it is at this writing. It is now known as Connecticut Aero Tech School.

CHAPTER FOUR

Getting Licensed

After graduating from Putnam Tech, I went back to work for Bert. My goal now was to get my private pilot's license and my FAA mechanics licenses. The private pilot's license permits the holder to take passengers on flights, but not for hire. The FAA airframe and powerplant (A&P) mechanics license entitles the holder to repair and inspect U.S. aircraft and certify the repairs.

THE DAY I NEARLY GOT KILLED

The minimum flight time to apply for a private pilot's license was forty hours, so I decided to fly one hour every noontime. I took my lunch in the Aeronca Champ at twelve sharp every noontime, and I had someone give me a start by hand-cranking the propeller.

When I had completed thirty-seven hours, I decided to practice stalls at two thousand feet. I throttled back and pulled the ship up in wing angle. I was watching the wing angle out of the left side when I realized it was awfully quiet. I looked forward and saw that the propeller was dead stopped; I was looking at its wood grain! The engine had quit!

I had been trained for this event, so I immediately picked out the biggest field around and headed for it, gliding. Now the engine was **windmilling**; it was turning over because of the air rushing through the propeller. As I started over the edge of the field, I could see that it was full of big boulders, which I had not seen from a distance. Just then the engine started running **full boost**. I had a split second to decide what to do. If I started across the field and the engine quit again, I would go into the trees on the other side for sure. If I tried to land the plane, I would likely snag a boulder, and the airplane and I would become a total loss. I clearly remember that, strangely, I was not scared whatsoever; rather, I was thinking what to do. My instructors had taught me to **spiral up and down**—pull into a tight turn and pitch up or down—so I pulled the Champ into a tight upward spiral. I figured that if the engine kept running, I might get high enough to reach a better field, and if it quit again, I could spiral down and be no worse off. The engine kept running, so I got up to about fifteen hundred feet, where I could see the airport. I headed that way, watching for other landing fields in case the engine quit again. The engine kept running, so I pulled into the pattern and throttled back. The engine quit again! Now I just concentrated on landing **dead stick** in the middle of the field.

When I stepped out of the plane, gasoline was running out. I turned the gas shutoff valve to off, and I walked across the field to the hangar.

Bert looked at me in surprise. "I didn't hear you pull up," he said.

By now, I was a little rattled. "The Champ quit on me," I said. "It is out in the middle of the field."

I briefly told everyone what had happened. We all walked out and pushed the Champ into the hangar.

"Harvey," Bert said, "take the carburetor apart."

I discovered that a small piece of rubber had gotten into the **float valve** and jammed it open. The engine ran okay at normal throttle, but when I throttled back, it loaded up with gas and quit. Problem found! Proof positive.

By the time I got the airplane back together, it was quitting time. I was tired and ready to go home. But Bert said, "Harvey, fly it around the field."

I was 100 percent sure that I had fixed the problem, and after having nearly been killed, I just did not feel like flying. But, of course, that was the point. Bert wanted me to get right back in the ship and fly it. That is, to get back on the horse that had bucked me. This is very important. You do not want to go away, obsessing about what might have happened. So I flew it around the field, tied it down, and went home.

Here is the kicker to this whole story: At age eighteen, I never thought to mention how, when the engine quit the first time, I'd decided to spiral up out of the boulder-filled field and thus probably saved the airplane and my life.

MY PRIVATE PILOT FLIGHT TEST

After I had completed the required hours for a private pilot's license, I took the written test and went for my flight test at Hiller Airport in Barre, Massachusetts, operated by Catherine Hiller. The date was August 11, 1954, about two weeks short of my nineteenth birthday. Catherine was a woman way ahead of her time. She was a 1990s woman back in the 1950s. She was well educated, and in addition to running the airport, she published the local newspaper. We did all her maintenance work, so I already knew her when I showed up for my test in the Aeronca Champ.

When we walked out to the airplane, she said in kind of a wry tone, "Well, Harvey, I know you are a licensed aircraft mechanic, so I will not ask you to show me how you do a preflight inspection."

We took off, and I did all the maneuvers she asked for—until the end, when she told me to do a wheel landing. A **wheel landing** is where you land on the main wheels first with the tail up in the air, as opposed to a **three-point landing**, using the tailwheel. It requires flying the airplane just off the ground, parallel to it, and then "**popping it on**." I thought I had done it pretty well. But . . .

She piped up and said, "I have got it. I will show you how it is done."

True to her words, she took the ship around and did a perfect wheel landing. She then went in and wrote out my license. I still think there was a little of the man-woman thing going on that day, but she did give me my license. I think that Catherine, as a woman in what was then totally a man's world, had a habit of showing her power now and then, in order to keep men in line. On the whole, though, I liked her.

MY FAA MECHANICS LICENSES

On July of 1954, I flew up to the Westfield, Massachusetts, FAA office and took the tests for the mechanics licenses. The runway there is nine thousand feet long and has a dip at the end. I landed right on the end, as we did at West Brookfield, but I could see *nothing* because of the crown in the runway! I took off again, flew up the runway to the first intersection, landed, and went into the FAA office. It turned out that I passed everything except the electrical part of the tests. I took that part over, passed, and was granted both licenses on September 9, 1954.

Each license has the FAA number you are assigned, and whenever you work on an airplane, you sign an aircraft logbook and specify that number. I had studied so hard to get my licenses that I clearly remember this number more than sixty years later, and I will remember it for life.

CHAPTER

FIVE

Working as a Certified Aircraft Mechanic

Once I was licensed, my work at West Brookfield Airport fell into a predictable routine of engine overhauls and aircraft inspections. After a few months, I was able to do a complete top overhaul on a 65 hp engine—the kind used in Piper Cubs—in a single day. I would pull the plane into the hangar at 8:00 a.m., remove the cowling and the cylinders, grind the valves, replace the rings, and reassemble and run the engine outside by 5:00 p.m. Aircraft inspection consisted of removing all inspection plates and checking the aircraft for cracks, corrosion, rust, or any other problems. There is also an FAA list of alerts for each type of plane to check for possible problems. They are called airworthiness directives (ADs) and they typically outline the problem and describe the specific action to be taken. Piper Cubs are very reliable airplanes, but they had about fifteen ADs, all of which were very minor, such as a nonserious cable rubbing situation.

Six-cylinder Continental aircraft engine. (Courtesy of Matthias Sieber)

The Temco-Riley Light Twin. At West Brookfield Airport,
we were kept busy maintaining one like this in first-class condition.
(Courtesy of Ken Stoltzfus)

THE BIPLANE GLIDER HITCH PROJECT

Catherine Hiller had a busy glider operation and decided to buy another tow plane. She bought a Fleet biplane that was a World War II Canadian trainer. The Fleet was a fabric-covered biplane with a five-cylinder engine. Its five cylinders gave the engine a distinct sound—something like a bulldozer.

I was handed a **glider hitch** and told to put it on the Fleet. I guess this was my first de facto engineering job. The hitch itself was just a piece of metal about two inches by three inches, with a **snap release** fitting to release a tow cable. But there was no place to put it! So after looking over the biplane structure awhile, I decided to cut the rudder away a little and then fabricate a boom to fit under the rudder and back into the fuselage structure. I cut away the fabric and welded in some bushings so that the boom could be bolted and unbolted from the aircraft. I fabricated everything and bolted it in place.

An FAA inspector showed up to examine what I had done. He looked at it, and to our amazement, he kicked it and said, "This is no good."

"What do you mean?" Bert asked.

"Well," answered the inspector, "to prove that it's safe, you'll have to block the wheels and pull on that hitch with a car or something like that."

Bert argued that that was not a good idea, that we might damage the airplane. But the inspector just got into his car and drove off, leaving the approval in limbo. We wondered what would happen, but a few days later, we got an approval letter.

In retrospect, now that I'm an engineer, I'm sure that if we had done what the inspector suggested, we could have pulled the airplane in two! The boom I had fabricated was about three times stronger than the fuselage.

The hitch went into service with no problems. I was later told that on one occasion, the Fleet even towed two gliders in tandem.

OTHER FIXES

One day, we were working on a North American Navion four-seater airplane with a bad magneto. A **magneto** supplies the spark on an aircraft engine. I pulled the mag off and worked on it all afternoon. But when I put it back, it still was not right.

It was the end of the day, and very uncharacteristically, Bert said, "Let's go for a ride." (An aircraft engine has two magnetos for safety purposes, but now the Navion had only one working mag.)

We got in and went over our house just as my dad was coming home from work, getting out of his car. My dad and Bert were friendly, so Bert said, "Let's give your dad a good welcome home."

He flew back off about a half mile away and laid on a good, old-fashioned **buzz job**. Our house was an 1800s Colonial with three chimneys, and we cleared them by about fifty feet.

When we got back to the airport, the phone in the hangar was ringing, so I answered it. A very agitated woman was complaining that an airplane of ours had come so low that all her chickens piled up in one corner, and some were dead. I finally recognized her voice as our neighbor, the one we never got along with. She knew I worked at the airport and undoubtedly hoped to get me in trouble. I quickly said, "When the pilot comes in, he will be reprimanded." For good measure, I added something else that I cannot remember. She was happy, and I hung up.

On another day, Bert had returned from a trip with a Navion and was standing by the **horizontal stabilizer**, pulling up and down on it.

"Harvey," he said, "go and get me a Phillips screwdriver." Then he said, "This thing does not fly right."

He used the screwdriver to remove a soft aluminum trim piece or fairing and discovered that the main aluminum fitting holding the tail on was broken! The only thing holding the tail on was that aluminum trim piece!

The FAA does a terrific job of notifying aircraft owners of air worthiness problems as they arise by mailing notices out immediately to maintenance shops. In all irony, we got a letter that very same day from the FAA to inspect all Navions for cracks in this fitting. We soon received a rework kit with a new, beefed-up fitting for mandatory installation. This was my first exposure to **fatigue stress**, which I would learn a lot more about as an engineer.

As mechanics, we often saw obvious areas where designs could have been made better and safer. In my opinion, doing mechanic work would make anyone a better engineer down the road. There is nothing like some hands-on experience with wrenches, rivet guns, and welding equipment as a precursor to designing airplanes or spacecraft equipment.

BUZZING BEARCAT

Life at our small airport was never dull. Occasionally, military or airline pilots whom we knew would buzz the field. One day, an F8F Bearcat with U.S. Navy markings came across the field one hundred feet off the ground at 450 miles per hour and three hundred feet from the hangar. This emptied the hangar of all people in a few seconds, due to the horrendous noise. The Bearcat had one of the largest reciprocating engines ever made: eighteen cylinders and 2,800 hp.

Bert generally knew many who were flying military aircraft in our vicinity, but this time he said, "I have no idea who that could be."

At that time, we had a salesman who called every month from Oilzum Oil, who provided high-quality oil for our aircraft engines. He was not a typical salesman! He was a very quiet gentleman who was always wearing a suit and tie. He would simply stop for two or three minutes, look at our oil supply himself to see what we needed, and go on his way.

On his next sales visit, though, he asked, "Did you see me last month?"

We found that question baffling. "Sure," we said. "We ordered some oil from you."

"No, I meant in the airplane."

"What airplane?" we asked.

"The Bearcat."

We were totally amazed! We had never even known that he knew how to fly. We learned he was in the U.S. Naval Air Reserve at South Weymouth, Massachusetts. He could fly all right!

GENERAL SESSUMS

I was in the hangar alone one day, working on a wing that we had stripped the fabric off of, so we could inspect the structure. A gentleman of fifty or so came around the corner and stood beside me, looking at the wing. He had a Hawaiian shirt on.

After a while, he said "There's the torsion box." I had no idea what he was talking about. (I learned later that the **torsion box** is the construction of ribs in the wing that absorbs torsion loads in flight.)

I knew this guy was someone special. About then, Bert came around the corner into the hangar and immediately snapped to. "Hello, General," he said in a very friendly manner. "How are you?"

It turned out that the visitor was none other than Major General John Walker Sessums Jr., who had been the head of Air Materiel Command at Wright-Patterson Air Force Base, Dayton, Ohio, and was now vice commander of Air Research and Development there. After a while, he started telling us about some of his projects. In his position, he could initiate and fund a project. He said he had been in the back of a C-54 Skymaster, coming back from some Pacific island, and he had started thinking about the fact that when an aviator is lost, it is not only a human loss but a loss of all the money spent on training the individual. General Sessums had an idea: Why not take the bailing pilot's inflatable life vest a step further and add inflatable wings and a put-put motor and propeller, with just enough oomph to get the pilot across enemy lines? He said he did have Goodyear build a flying prototype of his idea.

General Sessums was at our airport with his personal pilot, Francis Rohan, who had flown a Waco biplane at our airport before his military days.

THE PRATT-READ GLIDER PROJECT

One very interesting project at the airport involved replacing the exterior fabric of a World War II–surplus Pratt-Read glider. These gliders had been made during the war by a piano factory in Deep River, Connecticut. They held the world's altitude record for many years at forty-one thousand feet. It was towed up to ten thousand feet over the Cascade Mountains in California and released. Reportedly, the pilot was still going up in the updraft but had to come back down because he was running out of oxygen.

I would not really appreciate the aeronautical refinement of this design until years later. Modern sailplanes jam the pilot's feet and body into a pointy nose with little to no aerodynamic gain. With the Pratt-Read, though, two people can sit side by side in comfort. This glider should really be categorized as a **sailplane**, a term that denotes high performance.

We did have a problem during the re-covering process. Usually on airplanes, we would attach the new cotton fabric to the wing and wet it with water to pull the wrinkles out. We did this with the glider, but the wing construction was much flimsier than that of airplane wings; when the fabric started getting tight, it looked like the wing ribs might buckle or break from the stress. So Bert took a sharp knife and relieved the load by cutting the fabric along the trailing edge of both wings. We had to start all over with new fabric, but we avoided a major calamity!

AUTHORITY

In spite of my youth, no one ever questioned my mechanic work or my capability. I had a serious mien that was to stay with me always. At one point, the FAA required that all Piper Cub wing strut fittings be sawed off and replaced with a stronger design. That was four welds per airplane, which would literally hold the wings on. I did at least ten airplanes. One customer, Gordon Richards, had a welding business and stood there watching me weld one day. He allowed that I did a "solid job." He turned out to be very supportive and often told me, "Harvey, you tend right to business!" This comment came from the fact that whenever Bert was gone, I did watch over the operation.

Pratt-Read glider. (Courtesy of New England Air Museum)

THE DAY I SAWED A GLIDER IN HALF AND LATER REPAIRED IT

A man brought in a World War II Laister-Kauffman glider that had been damaged and "repaired" elsewhere, but the horizontal stabilizer and the wings were on different angles, and the glider did not fly right. The owner stood with Bert and me as we looked over the problem. I cut all the fabric away in the affected area of the fuselage. We could see that the previous repair had been botched, and the only way to fix it was to saw the fuselage in half, realign the structure, and reweld it together. So we disconnected the control cables, and I sawed the glider in half. We then slid the aft part back a few feet. The owner looked at his prized glider now in two halves. He tensed up, said excitedly, "I cannot watch this," and left! I got it back together, aligned and welded as good as new, and it flew properly again.

THE FLYING WELDER

Since I loved welding, I became the airport aircraft welder. It is the kind of skill you have to keep at or lose. An airport manager named Stan from a nearby airport had some welding to get done on an airplane, and he called Bert. Bert asked me to pack up some welding equipment and fly over to Stan's airport in the Aeronca Champ. The work was in a very difficult location under the belly of the ship, so I had to lie flat on my back and weld in a piece of structural tubing. Stan watched my welding like a hawk through his own welding glasses. It did not bother me, as I was confident in what I was doing.

One winter, to keep the business going, Bert had me make a single Piper Cub out of three wrecked ones by doing whatever welding was needed. One was crushed in the cockpit area and the wing fitting area, so I welded in all new tubing and wing fittings. I made one landing gear out of two crushed ones. In the spring, the plane was sold, and it was as solid as new. Years later, I was amazed, looking back at the responsibility I had assumed in welding on wing fittings and wing strut fittings, but I am confident I did a good job.

CHAPTER
SIX

Flying as a Licensed Pilot

UNCLE WILLIAM, FIGHTER PILOT

My uncle William Richmond Jr., the one who had been a World War II fighter pilot and squadron leader, was now working as the postmaster for North Wilbraham, Massachusetts. He told me he "would love to get airborne again." I told him I had the use of a Piper Cub as part of my employment, and if he wanted to, we could go flying after work some evening. Soon afterward, I picked him up at the post office.

When he got in the car, I asked him, "How is the post office business?"

"Boring as hell!" he replied.

We went off around his house in Wilbraham, with him flying. I will always cherish that flight. Shortly thereafter, he was coming home from parachute jumping at Orange, Massachusetts, went off the road, and died in his car.

A FEW HIGH JINKS WITH AIRPLANES

At one point, I was given the use of an 85 hp Cessna 120, which I flew often after work in the summer. The 120 is a two-seater, high-wing aircraft, all metal except for the fabric covering on the wings. Although I was supposed to stick to mainly straight and level flight, I decided to try some wingovers in the nice, calm air near sunset. These were nothing like what my flight instructor, Paul Cormier, could do, but I worked on getting them smooth. I enjoyed the kind of zero-gravity effect when the ship fell away at the top. I invited Bob, a student pilot, to go with me a few times. Of course, I did not mention my wingovers to Bert.

One day, Bert said in a friendly and congratulatory way, "You know, Bob told me, 'Boy, Harvey does those wingovers really nice.'" Oops! I was ready to get chewed out, but instead, he was pleased! Whew!

One customer, Jimmy, owned a newer two-seater, 90 hp Cessna. He was always very friendly, but he was constantly suggesting we should race the two planes to see which one was faster. Of course, he assumed that with a little more horsepower, his was faster. But I knew that our ship had smooth, fabric-covered wings, whereas his were sheet metal with rivets protruding and probably more **wing twist angle** than ours and therefore more **drag**. Finally, I relented, and we carefully planned the details of a race. We would fly to an isolated country spot, lock our throttles at cruising rpm, wave, and see what would happen.

I locked my throttle and started watching for Jimmy out of my left-side window. Soon he was falling away from me, and then he was out of sight. I broke away to the right and went back to the airport. I landed, and he landed soon after me.

He immediately stalked up to me—no more the friendly Jimmy I had always known, but now a very angry man. "You S.O.B.! You firewalled your throttle [went to full throttle], didn't you?"

I was astonished. "No," I answered emphatically, "I did just as we planned. I did not touch my throttle."

"Well, when I saw you going away, I firewalled mine." Jimmy was inconsolable for a while, but he eventually got over it!

Later, when I was working toward my aeronautical engineering degree at Indiana Technical College, I would learn that rivet heads cause what is called **parasitic drag**. Our ship had an adjustable wing twist, and we mechanics had set it for minimum drag. Sorry, Jimmy!

AIRPORT CHARACTERS

Small airports seem to attract more than their fair share of characters. One day, an older gentleman came in and quietly sat on the steps from the hangar to the office. He had been around before. After a while, he said, "You know, during Prohibition, they used this field to fly booze down from Canada at night in a biplane. . . . They would load up one cockpit."

I was surprised, as I had been around the airport for about five years by then and I thought I knew all about its history. "How do you know that?" I asked.

"I am the one who put out the smudge pots, so they could see to land!"

For years, another local gentleman would stop in and watch us work, puffing his pipe and saying nothing. We did not know anything about him. Finally, we learned that he had been an engineering officer on fighter airplanes during World War I, and he started telling us about some of the problems they had with rotary engines. A rotary engine is where the crankshaft is fastened to the airframe, and the whole engine goes around with the propeller attached! Eventually, he started showing up wearing a flight jacket with a World War I squadron patch on it. He was different, and his thoughts seemed to be far away. This went on for about a month, and we humored him for the most part. One day, his wife called to say that he had passed away. She thanked us for being so nice to him! Evidently, when he came in among us and our airplanes at the end of his life, he was reliving his World War I aviation days.

HOT ROD DAYS

Since I was now working full time and had a little money, I did what every red-blooded young man in those days did: I got a hot rod. I bought a 1932 Ford three-window coupe and spent winter evenings building up a souped-up Ford flathead engine for it. When the engine was done and installed, it was plenty fast!

The author with his 1932 Ford coupe hot rod.

A REALLY SHORT LANDING STRIP

One day, a gentleman came in wearing very thick glasses and politely asked in a French Canadian accent if there were any airplanes for sale. He said he had a five-hundred-foot strip up in Warren, Massachusetts. We were up to our hips in work, and Bert told him, "Sorry, no."

"Thank you," he very politely responded, and he left.

The next day, he appeared again. "Excuse me," he politely said. "After I left here yesterday, I bought a Piper Cub airplane at Leicester Airport, but it has a small problem. It has a leak about halfway up the gas tank." The Piper Cub's gas tank is right in with the pilot, so any leaks would be directly on the pilot. This is an extremely unsafe condition, so now he had our full attention. "Could you repair it?" he asked.

Again, we said we were too busy, and again he left.

After a few weeks, he reappeared and told us he was flying out of his field. We inquired about the gas leak, and he said, "Well, I just do not fill the tank up more than halfway."

After that, the airport owner, Gus Willet, told me, "I have got to see this." Gus mainly wanted to see the field.

So we got into his Aeronca Champ and found the airstrip. Sure enough, it was only five hundred feet long, with trees on both ends. As it turned out, the airplane was not there that day. Gus decided to land.

Now the problem was getting out of this strip. It was around suppertime, with no breeze to help with the takeoff. Gus sat with the engine idling, looking down the strip for the longest time. He was a superb pilot, and I would rely on his judgment any time. Finally, he started the Champ downwind. He got the tail up and swung the Champ around while on the run. Now we were airborne but looking at *very big* trees. At just the last moment, he turned the Champ on its side and sliced one wing through a slot in the trees.

When we got back, I said something to the effect that that was kind of close. Gus said, "Yes, I was thinking of asking you to step out, and I would come and get you with a car."

He had every step planned in advance, including, if we were not in the right position to make it, the **cutoff point**, the last point where he could cut the throttle and still have enough room to land.

CHAPTER SEVEN

Becoming an Engineer

We had some engineers who flew at the airport. From what I could gather, it seemed to me that engineering was the next step for me, so I decided to become an engineer. Families today plan college admission for their sons and daughters for years. I applied, was accepted, and quit my job all in one day!

Here is the exact sequence: It was about the last day of August 1955. I called Bert and told him that I would not be in and that I was going to see about going to college. I drove to Putnam Tech (half of which had been washed away by a devastating flood a couple of weeks earlier) to get a copy of my grades. I also saw Mr. Ellis. He had had a stroke, so they had another man actually running things, with Mr. Ellis as an honored advisor. I could tell Mr. Ellis was proud that I was going on to college. He shook my hand and wished me well.

I then proceeded to Worcester (Massachusetts) Junior College to apply for the mechanical engineering curriculum. I showed up with no appointment and was told I would have to see Dean John K. Elberfeld. He looked at my grades, kind of chuckled, and commented, "Well, these are the equivalent of all As and Bs. I have two seats left, you are in, and school starts Monday." He then gave me a list of books to buy.

I drove back to the airport and told Bert that I had to go to college, starting Monday. He was extremely unhappy, and in retrospect, I do not blame him. After all that he had done for me, I was giving him no notice. After a few days, though, Bert calmed down and told me I could work at the airport Saturdays and Sundays. For two years, that's what I did: I lived at home, drove every weekday to school, and worked every weekend at the airport.

Initially, I had a deep concern about how well I would do in college. Some of my friends had told me I was not prepared, particularly in mathematics. But I worked hard, and the first semester, I was on the dean's list. The curriculum was a good match for me.

After a couple of months, I sold my pride and joy, my baby blue 1932 Ford three-window coupe hot rod, to pay toward my college tuition and bills. It was one of the best investments I have ever made! Also, from working weekends at the airport, I had enough money to pay for the rest of my tuition and all other expenses.

Professor Pote, who taught our physics class, had worked on the Manhattan Project in World War II and certainly was a well-qualified teacher. He could be very no-nonsense and brusque, however. One day after class, another student and I approached him and asked innocently why it is not possible to exceed the speed of light. He dismissed us in our tracks with "It is impossible, period!"[3]

I became friendly with another mechanically minded student. We were in the Mechanical Engineering Lab, looking at an ancient

---

3. The speed of light is the next barrier to exploration of the universe. Humanity is always working on barriers, and frequently, they are overcome. A Mexican physicist, Miguel Alcubierre, has proposed a theory that meets Albert Einstein's equations and permits speeds greater than the speed of light. However, it would require generating massive amounts of negative energy, and the means are not known.

single-cylinder engine that had two three-foot-diameter flywheels on it from the early 1900s. We wondered if they ever ran it and asked the lab instructor about it. "That thing has not been run in twenty years," he said. "If you can get it running, I will give you each a full lab credit."

We brought in a battery, wire, and some gasoline and had the engine running in no time. Unfortunately, the college was in a multistory building, and vibrations were shaking the whole place. I was amazed when Dean Elberfeld himself appeared within thirty seconds. For an instant, I wondered if we were in trouble, but he was all smiles. He watched the engine run for quite a while and gave me a knowing look through the noise and vibrations, no doubt remembering my mechanics background. It turned out that Dean Elberfeld had written a textbook on metallurgy and taught the class on that subject. I attended, listened carefully, and learned a lot.

Worcester Junior College gave me mechanical engineering and physics basics at a cost I could afford. I received my associate's degree in mechanical engineering in June 1957.

CHAPTER
EIGHT

Marriage, the Marine Air Reserve, and Back to College

I met my wife, Cecile, on a blind date. I was unaware that she had previously picked me out of a group of budding mechanics at a garage in Brimfield, Massachusetts. She asked her brother, "Who is that skinny blond guy?"

"Oh," he said. "That's Harvey Smith."

"Where is he from? . . . I want to meet him."

So I was soon on a double date with Cecile and her brother and his date. She was sixteen at the time, and I was twenty. (At this writing, we have been married for sixty-one years.) Cecile was from a big family, with six brothers and one younger sister. Her folks were hardworking dairy farmers, and all the boys worked on the farm.

Cecile is a wonderful person! Well-mannered and very thoughtful of others. We dated for two years while I attended Worcester Junior College and worked weekends at the airport. Cecile graduated from high school as the valedictorian. On the same day, I graduated from Worcester Junior, so we could not attend each other's graduation.

We were married on May 17, 1958, at a small chapel at Brimfield, Massachusetts; the wedding ceremony was followed by a reception at her parents' home. Cecile's mother cooked up a storm, which was not unusual; they often had as many as forty people for dinner on Sunday because her relatives loved to "come out to the farm." The guests had to eat in shifts.

THE MARINE AIR RESERVE AND BECOMING A FATHER

Because all men had a military obligation at that time, I decided to join the Marine Air Reserve at South Weymouth, Massachusetts. They were flying F9F Panther jets at that time, and I thought some jet experience would be useful. The F9F Panther was the first successful carrier-based jet and had been used in the Korean War.

I did six months marine basic training at Parris Island, South Carolina, and the remaining three months at Camp Lejeune, North Carolina. Cecile was expecting our first child at that time, and she moved back home with her folks. Of course, our separation was difficult, but it had to be done. When I got home again, I went back to working for Bert and attending monthly drills at the U.S. Marine Air Wing at South Weymouth.

Our first child, Lori Sue, was born on February 23, 1959. We were staying with Cecile's parents, who had moved to a different farm in Warren, Massachusetts. Since it was February, I figured that when the time for the baby came, we would probably have a major snowstorm, so I had bought chains for the rear tires of the car. (For anyone unacquainted with these, they work!) When it came time to go to the hospital, however, it was a sunny, clear day, with no snow in sight; we just cruised down to the hospital. After we checked in, a very large nurse came out, pointed at the door, and told me, "Go home!" Which I did. This was the standard for fathers of that day; you were sent home!

BACK TO COLLEGE

I had worked on airplanes and flown them, and now my goal was to learn how to design airplanes. I applied to Indiana Technical College (now Indiana Institute of Technology) in Fort Wayne, Indiana, for the aeronautical engineering degree program. I was attracted to Indiana Tech because they offered an accelerated program with no summers off. I wanted to get it done! When the federal student loan program was enacted, I applied and was approved for a loan to cover my tuition for two years.

In late August 1959, with total resources of $300, a baby, and a positive attitude, we loaded up a U-Haul and moved to Indiana.

"Well, Harvey," one of the airport regulars, Gilson Merriam, said to me just before we left, "you got guts, but I think you are going to need something to put *in* your gut." He knew we did not have any money.

Soon, I got a job working for Professor Bennett Kemp in the wind tunnel after school and on Saturdays, doing welding and model preparation, which also led to work in the mechanical engineering laboratory, run by Dr. Ivan Planck and his assistant, Edmund Napier. Now I was splitting all my time after school between the wind tunnel and the mechanical engineering lab. Cecile worked as a carhop at Gardner's Drive-in Restaurant from 5:00 p.m. until midnight, outdoors, in all kinds of weather, including the Indiana winters. I would come home at five and take care of the kids and do homework while she worked. (Our first son, Edward, was born February 10, 1960. A second son, Alan, was born March 6, 1961; he would become better known to family and friends as Peter.)

The biggest problem for Cecile and me was money. *We had none*! We did not eat just beans, as I had at Putnam Tech, but I was trying to think of ways to stretch whatever money we did earn. One time, for example, I made a large pot of pea soup, which was very economical, and it seemed to last forever. We counted Cecile's tips on the kitchen table at midnight. We didn't become discouraged, though; we knew we were working toward a bright future.

On the first day of one of my aircraft classes, the instructor, William Hazard, asked, "Have any of you heard of an airplane called the Meyers MAC-145?"

I raised my hand because we had worked on one at West Brookfield Airport, and it was a terrific airplane for its time. It was pretty much a scaled-down World War II fighter aircraft with an excellent safety record.

Mr. Hazard then said, "Well, I was the co-designer."

He told us the plane had been built in southern Michigan by Meyers Aircraft, a company founded and still operated by a gentleman named Al Meyers. It was interesting to learn a lot of firsthand information about designing and building aircraft in a small class. For example, we learned about Mr. Meyers's test method for a new design: He put the plane in a terminal dive and pulled it out with the trim tab. A **terminal dive** is when you head straight down, and the airplane reaches a maximum speed based on its own drag. The **trim tab** is a small control on the elevator that can put a small load on it to very gradually pull the airplane out of the dive. Pretty dicey stuff!

I had come there to learn how to design airplanes, and I threw myself into the wind tunnel and airplane design courses. There were only six of us in our aircraft design class! I much prefer this to a school where class sizes are in the hundreds. In this class, we did all the calculations and drawings to design an airplane. I got the single A grade allowed.

The class was taught by Professor Siegfried Brunnenkant, a very enthusiastic teacher. He had emigrated from Germany, where he had worked on gliders. He had also graduated from Indiana Tech in aeronautical engineering and had designed a **flying wing** for his senior project. The flying-wing idea was very complicated aerodynamically and well ahead of its time. (On conventional airplanes, the tail counterbalances inherent wing rotation forces. On a flying wing, the wing is composed of several different airfoils; the different rotation forces achieve balance and eliminate the need for a tail.)

I wanted badly to take a rockets class that was offered, but there was a conflict: My chemistry lab overlapped the rockets class. I found that I could work like crazy in chemistry lab and get done in time to run across the campus to the rockets class without ever being late.

We lived in a school-owned house on the corner of the campus. One noontime, I came home for lunch and saw President John F. Kennedy on TV, saying that we had to get a man on the moon by the end of the decade, before 1970. "I believe that this nation should commit itself to achieving the goal, before the decade is out, of landing a man on the moon and returning him safely to the Earth." The date was May 25, 1961. The Russian *Sputnik* had scared the country silly. What is this thing? Could they put a bomb in it? I clearly remember those days! Kennedy's speech was a rallying cry for a country that was not feeling so good! What he left unsaid was that this new moon mission was also to restore the USA to the undisputed number-one position in the world. Little did I realize at the time that I would be working on this mission later.

A VERY INTERESTING AERONAUTICAL EXPERIMENT WITH A BIRD

One of the most interesting things I did at Indiana Tech was not in the curriculum. It was an experiment I did independently at our apartment with classmate Wilfred Beavin. We studied together sometimes, but at one point, we took a break. We started watching the caged finch someone had given Cecile.

Look at this, we noted. Here is an aircraft better than anything aeronautical engineers have come up with. *It reproduces itself!* We then wondered if the finch followed the same aircraft design rules we were studying.

First, we measured the **wing area**. We took the finch out of its cage, held one wing against graph paper, traced the area, and then counted the squares to get the wing area.

Now we were excited! If we could get its weight, we could calculate the **weight-to-wing-area ratio**, which is another key aircraft design characteristic. How could we do this with no scales? Then we got an idea. I had a three-foot yardstick and a round roller to serve as a pivot for a seesaw arrangement. We had a button whose weight we calculated by knowing the material, and using this arrangement, we determined the finch's weight.

Now, if we could determine the speed of the finch in flight, we could calculate its horsepower! We tied a light string to the finch and let it fly around in circles while we timed its trip and measured the circle's radius. We calculated the finch's minute horsepower; it was in seven decimal places, but we did get it!

Guess what! The weight-to-wing-area ratio was right in the range of modern aircraft. How about that? Smart bird!

GRADUATION

When I graduated in June 1961, we had just twenty-five dollars total in the bank for emergencies. I decided this was an emergency, so I bought a keg of beer at the local brewery for fifteen dollars and a big container of potato chips for ten dollars, and we sat out on the lawn with friends and had a good time.

Years later, it occurred to me that even when we barely had money enough for milk or food, I never once thought of quitting! It never entered my mind! When I told Cecile about that realization, she told me, "Well, I did think of it." But she never told me at the time.

CHAPTER
NINE

Working at a Start-Up

Now that I had my BS degree, our immediate need was money. There was an article in the paper about Telectro-Mek, a new company starting up in Fort Wayne with a system that measured jet engine thrust directly, which was intended to replace methods that used **engine pressure ratio (EPR)** as an indication of thrust. I got my résumé, put on a tie and sport coat, and rushed out there the same day.

I met with the president, Dan Russ, a very bright engineer and entrepreneur who had invented the thrust-measuring system and was getting government contracts to develop it.

It turned out that he was not only an engineer but also a pilot. He was interested in the fact that I was an A&P and a pilot because he was planning to buy an airplane in the near future. He had already hired two senior engineers, George Bertsche and Ray Harpel. George was a very experienced and creative electronics engineer, and Ray had been the supervisor of one hundred mechanical design people in a large corporation. I seemed to hit it off with everyone from the beginning.

During my interview, Ray asked me, "Why do you want to work here?"

"I need a job," I answered.

Ray laughed out loud with a hearty guffaw. He was very down-to-earth and appreciated my candor. They talked to Professor Ben Dow, the head of the Aeronautical Engineering Department at Indiana Tech, and I was hired.

So now I was getting a regular paycheck, but we decided to stay in the school apartment for a few months. When we had first come to Indiana Tech, we needed a soft chair, so I went to the Salvation Army and bought a soft swivel chair that looked good for three dollars. Unfortunately, when I got home, I discovered it would not sit straight but would always lean over. I could not afford to buy another one, and I was too busy to take it back. I swore that if I ever got a decent paycheck, I was going to burn it, and that is just what I did. In those days, everyone had a burning barrel, so one evening, I set the swivel chair on the burning barrel and lit it up. I was standing there enjoying the flames when my student neighbor, Angelo Boscolo, came running out. "Smitty," he said, "what are you doing?" He thought perhaps I was having some sort of a mental breakdown, but I explained the situation.

It turned out this job was a blessing. In a start-up company, you do everything there is to do—from sales to designing the product to testing it to putting it into a box and shipping it.

Ray told me I would start by making a layout of a **pneumatic computer**. I said, "What is a layout?"

Ray took a long pause, saying nothing for quite a while. Now, years later, I realize what he must have been thinking. Training someone to make design layouts is a task that typically requires years. But after a while, he started teaching me how to make a design layout.

The everyday work at Telectro-Mek was filled with a can-do spirit and creativity. It was a positive atmosphere of "why not?" which would stick with me. I worked hard and was eager to learn. I could see this was another opportunity.

We made prototype models of the company's thrust-measurement system for the air force, the navy, and FAA all at the same time. We made both pneumatic and electronic prototypes of the proposed **thrustmeter**. Pneumatic computers were strong in the technical literature at this time; they would be unaffected by nuclear radiation. My aeronautical engineering degree came in handy; I used my jet engine flow calculations from my Indiana Tech courses.

I was sent to the Naval Air Turbine Test Station in Trenton, New Jersey, for one month to monitor the testing of the Telectro-Mek probes in jet engines and the taking of data for thrustmeter use. The main issue was **probe survival**, particularly under **afterburner blast**, which exerts a terrific increase in pressure and temperature on the probes. We were successful in providing a probe that would survive.

I learned that Telectro-Mek received a government bid list every day called the Commerce Business Daily, which contained some aerospace-related jobs that could be bid on. I saw a request for proposal (RFP) for some contaminated-fuel detectors, ground test equipment to test fuel. I asked Dan, Ray, and George if they would like me to get the info needed to bid. They told me to get all the costs together for the parts to arrive at a bid price. I put in every nut and washer, but I could not find a price for an electrical display meter that was specified. George, the electrical expert, told me to put in ten dollars. We submitted the bid and got the contract, only to find that the meters actually cost twenty-six dollars each. We needed one hundred of them, and the total cost was a big problem in 1961. We needed to find a way to bring down our cost. One of the main cost items was the case, or housing, for the instrument. Ray solicited bids from several fabricators and achieved major cost savings. We were able to produce the contaminated-fuel detectors successfully, at a profit.

Ray showed me how to make up design schedules with the two draftsmen I was supervising and to meet those schedules so that money would not be wasted. He also showed me how to be firm and insistent when necessary. He taught me how to confront people when I needed to and to bring things up to the next-level supervisor.

FLYING ADVENTURES

At one point, Dan told me he was ready to buy a plane and asked me if I would go with him by car on a couple of Saturdays and evenings to look at airplanes. After looking at several, we found a really nice Piper Comanche 160, very clean and in good shape. It is a four-seater aircraft with retractable landing gear and a good choice for his company use. Dan bought it and started keeping it at Baer Field in Fort Wayne.

It got to be routine that I would be working on my design board, and my phone would ring. "Harvey," Dan would say, "do you want to go flying?" Of course, it was a rhetorical question. I was going.

Usually we went to downtown Chicago, where Dan had business connections. We landed at Meigs Field, which was right downtown, out in Lake Michigan. I was introduced as either the copilot or the engineer, depending on the occasion. I usually flew the plane home while Dan slept because he was always worn out from his high-energy meeting performances.

One time, Dan asked me to fly up to Meigs Field and check out a Twin Beech Model 18 aircraft that he was thinking of buying. The Beech

18 would have been a big jump up for both of us, from a piloting point of view. It is a six-passenger, retractable-gear aircraft with two 450 hp Pratt & Whitney engines. The plane was also used by the U.S. Air Force, where it was designated as a C-45.

Although I had graduated from Indiana Tech, I was still a member of the school's Flying Club, and I used their Cessna 140 now and then. I decided to take it on this errand. I flew up okay, but when I went in on the final approach at about one hundred feet of altitude, a violent wind gust off Lake Michigan took me straight up about three hundred feet and then back down to fifty feet, regardless of what I did with the controls. This was the only time I felt completely out of control while flying. I was somewhat shaken, but I landed okay.

I checked out the Beech 18, including a detailed review of the logbooks. When I got back, I advised Dan against buying the plane because it had **high-time engines** that would soon require a very costly overhaul. But it had been an exciting trip!

Cecile and I could afford to do a little flying on the weekends, so I got checked out on Piper Cherokees at Baer Field. The Piper Cherokee is a low-wing, four-seater, all-metal aircraft with a fixed landing gear. We would all go for a local flight around the area with the three kids in the back. We never did fly to distant locations cross-country, due to the expense.

We did fly for a great many weekends in a row, however, because I was enjoying flying again after a long pause. The part of Indiana we were flying over was as flat as a billiards table. One interesting feature of the area is that all the main highways were laid out exactly north and south and east and west, so a pilot could set his or her **directional gyro** by lining up with one of the main roads and resetting the gyro to update it.

While Cecile was flying with me one day with the kids in the back, a thought occurred to her: What do I do if he gets incapacitated while we are up here? She decided to take flying lessons at Baer Field, earning all the money to do so herself.

When it came time to solo, she was six months expecting. "Mrs. Smith," the instructor said, "you either have to solo today or wait until the baby comes, because you have to be able to get the wheel all the way back."

"I'll go today," she said.

She did solo, and she did an excellent job. It was a Saturday, and all the Air National Guard jets were flying in a crowded sky. My wife is probably the only woman in the world to solo an airplane while six months expecting and who earned her own money to do it. (Dwight was born July 23, 1963.)

LEAVING TELECTRO-MEK

I was very happy at Telectro-Mek, but my dad had a heart attack and Cecile wanted to go back home, so in the autumn of 1963, we moved back to Massachusetts.

The day I left, Mrs. Russ told me, "Harvey, I always thought it was so enterprising of you, the way you came out here."

I had learned a lot. Most important, I learned what a positive attitude will accomplish.

CHAPTER

TEN

The Apollo Program

In the early fall of 1963, I went to work on Project Apollo, the Apollo moon landing program, at the Hamilton Standard division of United Aircraft Corporation in Windsor Locks, Connecticut.

We had come home on vacation in September 1963, and before leaving Indiana, I made interview appointments at Raytheon Missile Systems Division, in Bedford, Massachusetts, and at Hamilton Standard, Space and Life Department, in Windsor Locks. The interview at Raytheon seemed to go okay, although I saw only one person, a group manager. At Hamilton Standard, my father-in-law went with me and waited in the car. He had a vested interest; he wanted Cecile and the kids living near him. The interview at Hamilton was intense. Phil Gaffney, whom I would later work for, actually asked me to help solve a design problem he was working on, and I gave him a few ideas. I got offers with exactly the same salary from both firms, and I chose Hamilton because it was a lot closer to where we wanted to live.

Working on the Apollo moon landing program was very exciting! This was not like going to work in an automobile or aircraft factory. Everything was an invention. Many times, there were TV crews at Hamilton out front or in the building. When I got home, Hamilton was often on TV. There was something special in the air. This was a national priority! Soviet cosmonaut Yuri Gagarin had been the first person to orbit the Earth on April 12, 1961. Clearly, the United States was in second place. We needed to catch up!

I was hired as a design engineer. Design engineers take ideas from scientists and turn them into reality, using metal, bolts and nuts, rivets, and plastic. For an engineer working on the moon program, this was the most interesting work anywhere. Traveling to and from work and at odd moments on weekends, I would think about how to solve some design problem. I would often be arriving home in Massachusetts when a beautiful full moon was rising over Minnechaug Mountain in the east. There was our goal right in front of me. Such a challenge. Could we get there? Would we get there? It was clear that the whole undertaking was gigantic!

All design engineers at Hamilton worked on a drawing board and used a **slide rule**. Slide rules performed multiplication, division, trigonometry, and logarithmic functions. We did not use computers. I am convinced we did just as well because we tested everything for strength and functionality. The only computer on the Apollo program was the

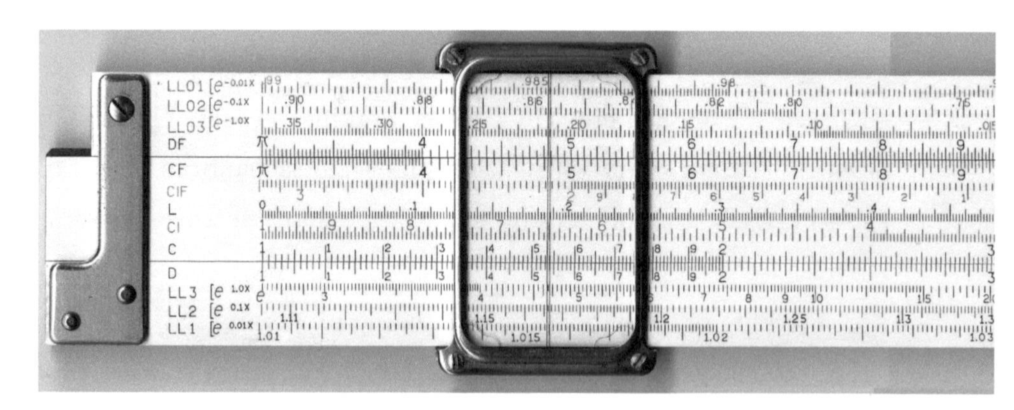

The slide rule, the way we got to the moon.

guidance computer used to guide the rocket to the moon and back, which was roughly the size of a home computer today.

We all wore white shirts and black ties. I do not know why exactly. It was not written down or posted. We just did. Also we mostly wore crew cuts.

I walked into work with a brisk step each morning, eager to get going for another day of excitement. Hamilton had been awarded contracts for the **environmental control system (ECS)** for the **lunar module** (**LM**, pronounced "lem") and the space suit backpack, known as the **portable life support system (PLSS)**. Hamilton had subcontracted the space suit to ILC of Delaware. Hamilton was a good choice to build the LM ECS and backpack because they had years of experience building aircraft pressurization systems. For example, Hamilton would build the Boeing 737 and 747 pressurization systems.

The LM ECS and backpack would provide the astronauts with the same partial oxygen pressure as here on Earth. In October 1963, Hamilton had already built a prototype backpack, and it was pictured on an ILC training suit on the front cover of *Aviation Week & Space Technology*.

I was initially assigned to work on the LM ECS, and one of the first

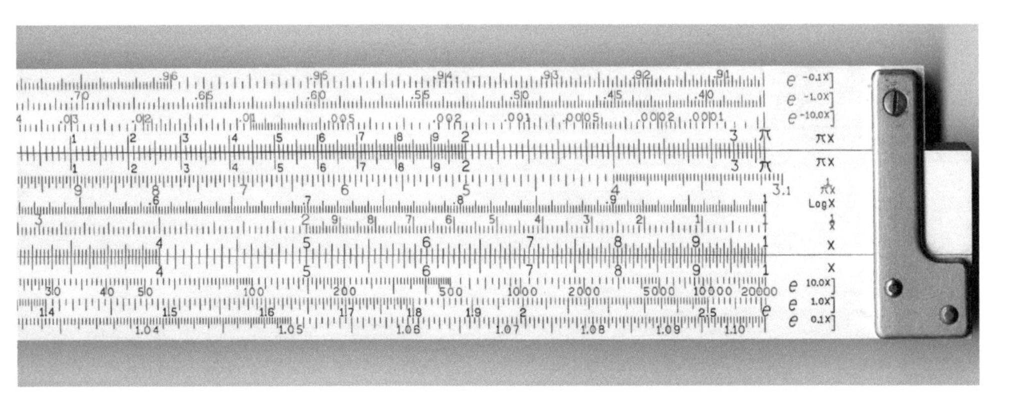

Aviation Week
& Space Technology
75 Cents
A McGraw-Hill Publication

October 28, 1963

SPECIAL REPORT:

Apollo Spacesuit
Design Program

Apollo Suit Prototype
With Life Support Pack

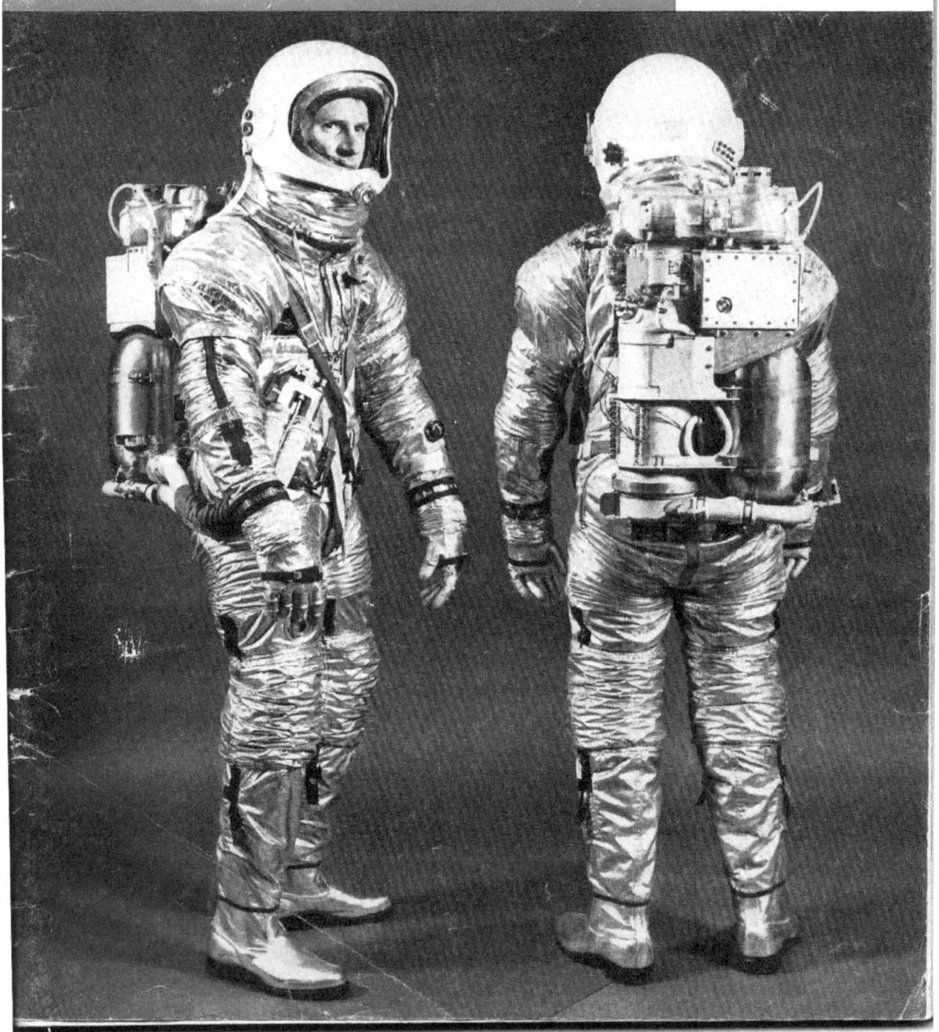

The prototype Apollo backpack on the cover of the October 28, 1963, issue of *Aviation Week & Space Technology.*

things I did was to go to Grumman Aircraft Engineering Corporation (later Grumman Aerospace Corporation) in Bethpage, on Long Island, New York, the prime contractor for the LM, so I could see a full-scale mockup. After walking up next to it, I was surprised that it was so big.

At that point, Hamilton was going to supply some water tanks, so I was assigned to start working on how to mount them and how to measure the water in them under **zero-g conditions**. Some of the tank design problems for this task were that in zero gravity, water floats around in gobs. Also, the rocket produces what is called **random vibration**, vibration that does not follow any predictable pattern and was a new kind of load not well understood at the time, which required testing. I did come up with one idea, a design for the water-measurement problem, which consisted of a bladder with metal hoops to control the water and a spring-loaded **potentiometer** readout. The design was deemed patentable.

Plans changed, though, and Grumman decided that they would be responsible for the water tanks.

DESIGNING FOR SPACE

Designing for people going into space brings on a whole host of new problems:

First, when two bare metals meet in the vacuum of space, they weld together, so all metals that are in contact must have a protective coating.

Second, plastics outgas, giving off terrible odors that humans cannot stand. In response to this odor problem, Hamilton set up a Smell Panel, a group of people who would meet weekly and smell different nonmetallic materials that had been subjected to reduced pressures, such as those the space suit would be subjected to. My job was to select only "smell" approved materials for design. It turned out that we needed to eliminate most common plastics and paints.

Third, space equipment requires special blankets to protect it from **micrometeorites** and radiation. It was the job of the Mechanical Design Group to see that blankets were designed to meet the thermal and meteorite requirements. The blankets had to stop a .22 caliber bullet.

Fourth, to enable an astronaut to walk on the moon in a space suit, Hamilton needed to devise a means of removing CO_2 from the suit. Lithium hydroxide had been used in submarines for years to remove CO_2, but it has the consistency of kitty litter. This material would now have to withstand rocket vibration. Filter and containment materials were developed that would allow lithium hydroxide cartridges to meet vibration and breathing requirements. Our group reviewed and incorporated the results of the Project Group's development efforts.

Fifth, Hamilton needed to devise a way to get rid of the astronaut's body heat in an airless environment, another major hurdle to walking on the moon. To solve this problem, the company invented the **porous plate sublimator**. This key invention involved feeding water onto one side of a plate with tiny holes in it; on the other side of the plate is deep space. The water freezes and then sublimates on the deep-space side. Water is continuously resupplied on the interior side, providing a cooling effect that is used to reject the astronaut's work energy into space.

Sixth, since any spark in pure oxygen would be deadly, we needed brushless motors to operate fans that would circulate the oxygen through the lithium hydroxide safely. We contracted with vendors to develop the motors.

A BIG GOOF AND A NEW WORD: INTERFACE

One day, we were all interrupted by a Grumman employee on the phone to his office at Grumman. He was waving a drawing and talking in loud, exasperated tones. A **fit check** mockup of the Hamilton LM ECS had been built, but it would not go through the door of the LM! This was a big goof of the worst and most unforgivable kind!

A new word that I would now hear a lot was **interface**, meaning the mating details of the mechanical and electrical connection between different spacecraft parts. The details can be from vendor to customer or for internal use. The problem is that humans are not perfect. In this case, someone had not submitted the correct interface drawing to the Hamilton designer. The package had been designed as rectangular, but the door was triangular! As simple as that. In the end, the Hamilton ECS package was redesigned to fit through the triangular door.

PRESIDENT KENNEDY IS ASSASSINATED

President Kennedy was assassinated on November 22, 1963. When it happened, there was a subdued murmur going around the design area, with people talking in hushed tones. When I asked what was going on, a supervisor cautiously said, "We have received a report that President Kennedy has been shot." He was careful to say it was only a report and not fact yet.

I reflected on the fact that Kennedy had initiated the work I was now doing, and I remembered how I had watched his famous speech that had changed history. It was tragic that he never saw the United States get to the moon!

THE SURVEYOR AND THE AHA MOMENT!

At Hamilton, we were designing things to go to the moon without even knowing what was there; we were designing to land a vehicle on a surface that was totally unknown! In 1963, there were two theories about the moon's surface: According to one theory, there was twenty feet of dust; the other theory held that the surface was as we know it to be now. In its original concept, the LM would have a hatch on top because the astronauts might have to exit out the top if the dust theory were correct. In 1966, when the NASA Surveyor program landed vehicles on the moon equipped with digger scoops and soil analysis equipment, we took a giant step forward. We all watched the videos and said, "Aha!"

The *Surveyor 1* footpad, which showed for the first time that the moon's surface was similar to Earth's. (NASA photo)

THE MODEL A PRINCIPLE

After working on water tanks, my next assignment was to design a valve that the astronauts could turn for more or less heat in their space suit while they were in the LM. The valve would be mounted on the wall in back of them. The astronauts would stand up for the moon landing, and would wear what was essentially a pair of long johns, with small tubes sewn in to allow fluid to circulate to keep them warmer or cooler as they wished. The fluid would come from what is called a **heat exchanger**, which is similar to a car radiator.

I was given a performance graph for the heat exchanger, which showed that it needed only a little bit of fluid to work initially but large amounts thereafter. We needed a straight-line increase in temperature to keep the astronauts comfortable. I came up with the idea of burning the exact profile needed on the graph into the inner moving part—a small opening initially with more afterward. Long story short, I used the Model A principle of simplest is best. My design did a complicated function with a single moving part! It worked like a charm and followed the flow objectives perfectly!

RULES AND PROCEDURES

I did have a problem adjusting to being a Hamilton designer because of the rules. I was used to calling vendors to get prices on materials. One day, a purchasing agent showed up at my desk with a red face and waving papers.

"What the hell are you doing calling vendors?" he demanded.

So I learned that calling vendors was a definite no-no. According to Hamilton company policy, like that of many others, only purchasing agents call vendors.

I was also used to going into the machine shop to see how my design was going. Unfortunately, that was a no-no as well. I quit calling vendors, but I would still make quick trips to the machine shop to see that the parts I had designed could be easily machined. I still think this is not only a good idea but also a necessity for all designers.

One thing that was confusing was security. The work we were doing was classified as confidential, which meant I could not tell anyone about it. Yet I would come home, and *Aviation Week & Space Technology* would have an article with all the technical details of our Apollo moon landing program. Now my dilemma was: Can I show this to my wife, show her what I am working on, without feeling that I am going to get arrested? Even more peculiar was the case where I had submitted a proposed patent for an emergency space suit helmet that would be sewn into a flight suit but would self-deploy if there was a sudden cabin pressure loss. I intended it for the early stages of testing. I was called to the security office and told that this had to be a "secret patent." It was never built, but the company thought it was a good idea at the time and wanted patent protection, which they got. The paper I signed essentially said that I should expect to be shot at daybreak if I breathed a word of this to anyone. Then, about a year later, I got a copy of the patent, released as a regular patent in the U.S. mail. Hello, can I show this to anyone?

Every designer at Hamilton had a BS degree in engineering. At the start of each design, there would be a concept meeting to ensure that the concept was right before we went down the wrong path. Next, the designer made a hand-drawn layout of the item, with enough views to completely define the design. This layout usually covered the whole drafting board.

We did **stress calculations** on the item itself for rocket vibration loads. We did a stress calculation and safety factor for each and every individual piece of metal, nut, and washer. A staff metallurgist reviewed all designs for material compatibility, strength, machinability, and plating. After all the reviews and sign-offs, the item was built.

Next, the component went for **functional testing**. For example, the valve I had designed was tested with a heat exchanger, just as it would be used in the LM. My supervisor came up one day with test results that showed that the valve worked perfectly, and he was very happy. My valve was considered very creative!

The next thing I knew, I was called into a meeting where I was told I would be part of a new team, tasked with designing and building a space suit for a new competition at NASA: the space suit to go to the moon! I was excited to have been chosen to work on building space suits. For an engineer, there was no better challenge anywhere!

CHAPTER ELEVEN

Of Space Suits and Helmets

I was assigned to work for a senior engineer named James Wilber on developing improved space suit joints, aimed at obtaining an Apollo space suit contract.

A major problem with suits at the time was mobility. Enabling an astronaut to walk on the moon and/or to move his arms and legs full range with ease when pressurized in a space vehicle was a new requirement. Although space suits are made of fabric, when pressurized to 3.7 psi, they become as "stiff as a pickle barrel," thus requiring flexible sections, or "joints," corresponding to the locations of the astronaut's body.

This was an exceedingly difficult problem to overcome. We learned that we could not design pressure suits like we did hardware. We had problems making progress, because the suits had to fit human beings of all different sizes, they had to be comfortable, and they had to have joints that moved with very low force when the suit was pressurized.

We had never even seen a **pressure suit**. We were starting from square one. We studied the work done by Litton Industries on an all-metal suit. We could see that the forces to move the joints on that suit were low, but there would be storage disadvantages, since the suit was hard and could not be folded up. That would be a penalty.

We also looked at training suits by B.F. Goodrich, David Clark Company, and ILC. We traveled to B.F. Goodrich in Akron, Ohio, to look at their pressure suits. Both their suits and the Clark suit were aircraft pressure suits with legs sewn in the position—which, of course, would be a limitation for walking on the moon.

Everyone working on suits was cleared by a medical doctor and scheduled to get in one and test it; the suit would then be pressurized, and the tester would try moving his arms and legs. An ILC training suit, which pretty much represented the state of the art at the time, was used for these trials. The trials made it clear that we had a long way to go before the suit we were designing would have the mobility needed to enable the astronaut to exit the LM and walk on the moon. There was no substitute for getting into a suit and feeling how limited it was in motion and how uncomfortable it was. In some cases, suit test people were actually shedding blood, due to poor fit. The ILC training suit had good elbow, knee, and shoulder mobility, but it had no hip joint.

I did a thorough literature search on space suits, including looking into what the Russians had done. In 1963, NASA published an internal report on Russian space suit technology, covering all aspects of suit mobility, **CO_2 removal system** tradeoffs, and reentry concepts. Also, a 1964 article published in a foreign news publication showed pictures of a spacewalk by Alexei Leonov. Our B.F. Goodrich associates at the time told us that under some government program they had lent one of their suits to the Russians and it had come back a year later in tatters from extensive examination. Evidently, they were no further along than anyone else at that time.

Hamilton had hired a PhD physiologist who acquainted us with anthropometric data on men and the huge variations in size from the 5th to the 95th percentile. Not only size but the position of the head relative to the spine. We finally decided that each astronaut needed a custom-made space suit; we would make a plaster cast of each astronaut and from that make a foam likeness.

We soon had designed and tested seven iterations of hip joints and five iterations of shoulder joints, but with no success. After all this effort over a period of several months, we had not even achieved something as good as the ILC shoulder, which had low torque and was relatively comfortable.

This was a very frustrating period. We would sketch up and build a joint idea that looked good on paper but failed to perform. Finally, Mark Baker's group took over joint development and started designing by an **iterative (cut-and-try) method**, based on what had been learned to date. Design changes would now be made "on the fly" with patterns only, bypassing the need for classic drawing-board design.

From then on, my group of designers documented all patterns for NASA and did various design studies for them. In one of these studies, we designed and fabricated a set of miniature fingertip lights to help the astronauts select switches under low light conditions. Also, Hamilton formed a business relationship with B.F. Goodrich, which would supply test suits and a shoulder joint.

Mark Baker did command a lot of respect for his common sense and his ability to get things done. At ten o'clock one night, I was just ready to go to bed when the phone rang. There was no "Hello, this is so and so." The voice on the other end said, "Harvey, get your ass in here! We have to run this test tonight! We have work to do tomorrow." He was referring to a test we needed to run on a pressure suit, where we would pressurize it and take measurements.

I got in the car and drove in, and we worked until somewhere around 3:00 a.m. I then went back home for a little sleep—and back to work by 8:00 a.m. The curious thing was that Mark really did not seem to think that running around in the middle of the night, testing a suit, was work! One engineer in the group worked forty hours straight that week, one hundred hours total.

The stress of these hours and the national urgency we were under created a difficult home life for all of us. Cecile was very supportive, and I did a little touch football with the boys when I could. (We did manage to take a family camping trip in Vermont with our Shasta camper and two tents for a few weekends, which was a great relaxer!)

My job was to manage the Mechanical Design Group, which did various design studies directly for NASA through a Hamilton project engineer. We also were charged with documenting all the patterns generated by the Project Group in the suit room, a lot of the work just copying shapes from rough patterns onto drafting paper to make a permanent record for NASA. Most of this work was done by contract designers, hired hourly through temporary engineering firms.

Two design studies we did for NASA involved control of feces and vomit in a space suit, politely called "waste management." (Urine control provisions had already been developed.) We first consulted with a doctor hired by the Space Department for this purpose. We told him we were considering some sort of strap-on design with a seal. He told us some lotion would be needed to avoid skin burn, as acids are involved. I assigned a contract designer to come up with some sketches of a molded seal with a bag attached, along with straps to hold it to the astronaut. To show the design, it was necessary to draw a man's butt with the proposed assembly held in place with straps. After a while, I noticed the designer was leaning over his board, quietly laughing. I went over to see what was going on. "When I used to do this in school," he told me, "the nuns would whack the hell out of me. And now I am getting paid for it!"

At home, I told Cecile what we were working on and how we had not come up with anything we liked. My wife, the mother of four, said simply, "Put a diaper on him." As history has since shown, that was the most practical solution for men and women astronauts. In fact, the Apollo space suit had a fecal containment system (FCS) as standard equipment. It was a large pull-up-style diaper.

We also studied vomitus, a very serious issue. First, if an astronaut were to throw up in the suit, he would be blinded because the inside of his helmet would be covered. If he were on the moon alone, this would be a big problem. Second, the vomit would clog up the backpack fan, and there would then be no oxygen circulation. We started brainstorming various ideas, but we never found anything that we really liked. Again, we were told that we *had* to design something!

One manager insisted that we consider channeling the vomitus through the water fitting in the helmet, which was something like a five-eighths-inch-diameter fitting. This would be like handing someone who was about to throw up a piece of pipe with a hole in it the size of a dime and expecting him to vomit through it. I told him flatly that I was not going to do it, that it was obviously a dumb idea and would be a waste of taxpayer money. He promptly went to my superiors and complained that Harvey Smith was not cooperating!

Still, I never had to do it. With my back to the wall, I did come up with an idea that I thought might have a chance. It involved placing a mouthpiece placed inside the helmet to the side of the normal space but accessible if the astronaut were sick. It would have a sewn-up bellows all folded up under the mouthpiece to hold the vomit. I made up some sketches and brought them into the sewing room for fabrication. The next day, one of the sewing ladies appeared in front of me and very bluntly asked, "Did you design this thing?"

I was a little startled because the sewing ladies did not often confront us. I stated that I had designed what she was making, and I asked, "Why?"

"This will never work," she said.

"Why not?"

"Because," she explained, "when I come home drunk and throw up, it goes all over the place. I could never hit this thing."

On reflection, I realized that she was right! I dropped the whole idea. I understand that vomitus control was ultimately addressed with diet.

We worked in a large secured room. Security required that you had to knock on the door and request entry. Your name had to be on a list, and you had to be recognized by someone inside. In spite of the best plans, however, things can always go awry. One time, the janitor opened one of the secure doors to mop up, and someone giving a factory tour to some junior high students said, "Oh look, kids! The suit room is open." He had them all inside, looking at all of the classified parts, before anyone could say a word!

I now had learned how to manage up to twenty designers and meet schedules. I knew that this would put me in a better position for promotion. About this time, I realized that I could either sit down and design something or I could manage a large group of people. At that time, the way to get promoted and earn more income was to become a manager, so that was the direction I headed in.

The Hamilton Standard space suit was not chosen by NASA in the competition. ILC of Delaware had solved its mobility problem and made the suit that NASA chose. They were awarded contracts and would make all the suits to go to the moon.

APOLLO HELMET DEVELOPMENT

Although I personally did not work on space suit helmet development, I was privy to watching the bubble helmet emerge that was used on the Apollo program. In 1963, initial testing was done with aircraft-pressure-suit-style helmets with movable visors. It became clear that a greater range of visibility would be desirable. Industrial designers were hired to try to "come up with something." One rendering had small decorative wings on the helmet, suggestive of the mythological god Mercury. None of these ideas caught on, but no one—including management—could describe what the helmet should look like. The Space Department had gone to the Plastics Department, who started working on blowing half bubbles out of sheet Lexan in an oven, but they did not produce a single helmet after something like a dozen tries. I attended a meeting where they reported that they had a perfect bubble when it "blew" (burst) unexpectedly. Finally, on the very last sheet of material, they achieved a half bubble, although on further examination, the optics were not usable.

Hamilton had hired a very skilled technician who could do wonders with either metal, plastic, or cloth. He quietly proceeded to blow a full, one-piece bubble helmet and put it on Vice President Edmund Marshall's desk. Finally, everyone to a man agreed! This is what a space helmet should look like! Personally, I think our enthusiastic endorsement came from the fact that people of our age group had grown up reading Buck Rogers comics from the early twentieth century, which always showed a bubble helmet. Nothing else looked right until the bubble came along!

Unfortunately, the full helmet produced by the blown method also did not meet optical requirements. Another firm, Air-Lock of Milford, Connecticut, started molding the helmets out of Lexan, and theirs was the helmet NASA chose to go to the moon.

CHAPTER TWELVE

The MOL Suit and the MOL ECS

THE MANNED ORBITING LABORATORY SUIT

After we lost the Apollo suit competition, management decided to fund a small group to try to secure the **manned orbiting laboratory (MOL)** suit contract. The MOL was a classified U.S. Air Force space program. We all guessed that it involved photographic equipment. We did have MOL astronauts come in to monitor our progress but under almost clandestine conditions compared with the Apollo program. Security was at a higher level on the MOL program, in that there was some secret information. There was no fanfare, and the program was never written up in trade magazines.

We were trying to get a suit contract based on what we had learned from the Apollo competition. A senior manager named Paul Stein was initially put in charge, and he was very good at getting the most out of small budgets. A new suit was fabricated by the Project Group, using the best joint knowledge gained on Apollo to date.

One of the details that came under my jurisdiction was once again to develop a "feces management" device. We restarted work on a seal and a

A MOL astronaut training in the Hamilton Standard suit. (USAF declassified photo)

bag. Soon, it was time for someone to sit in a bucket of wet plaster to get a shape for the mold. Time went by, and no one sat in it. Paul came over for a meeting one day and asked if the plaster mold was done, to which we reluctantly said no.

"Why not?" he asked.

"Uhh," he was told. "Nobody wants to do it."

He exploded: "You get that goddamn bucket ready, and I will be over myself to sit in it!"

It turned out that he never had to come over, because plans changed, and we no longer had to provide anything. A stop-work order was issued on the feces management project.

In May 1966, our suit was ready, and as the Mechanical Design Group leader, I wrote a test report showing the amount of force needed to move throughout the ranges of motion for all joints. The suit was found to have excellent mobility. After an evaluation by the air force, those of us in the trenches were told that although our suit had excellent mobility, we were not picked, because Hamilton was not a proven supplier of pressure suits. By "proven," the inference was that other potential suppliers, such as David Clark and B.F. Goodrich, had years of experience supplying pressure suits for aircraft use, and we did not.

Hamilton Standard did deliver seventeen training suits, however, designated MH-7 by the air force.

THE MANNED ORBITING LABORATORY ENVIRONMENTAL CONTROL SYSTEM

I was promoted to assistant design project manager and put to work under Phil Gaffney, coordinating one hundred engineers and draftsmen in six groups. Phil had been a World War II B-17 pilot, and I got along fine with him and worked hard. We worked in a classified secure room, where I attended morning meetings with Phil and the top managers. It was my job to go out into the unsecured office area and get everything done that required mechanical design action, whether initiated, communicated, transferred, or drawn.

The design starting point for the MOL ECS was the LM ECS, which had already been developed. The prime contractor for the MOL program was Douglas Aircraft. Hamilton had offered the completed LM package to Douglas as the most cost-effective approach, but it did not fit conveniently into the MOL. Douglas decided to buy the package but in pieces that they could spread around the MOL vehicle where they would fit best, interconnecting with the same environmental control results. Interface control drawings were made of all individual components and given to Douglas. I was told that the packaging was done by a single designer, working at a drawing board walled off with black curtains for security. James Bond stuff!

Initially, my task was to get the design program moving. It was a chicken-and-egg situation: The Systems Group needed to write specifications for the system, but to do so, they needed the Analysis Group to do some analysis, but Analysis needed to know what to analyze.

My job consisted of running around between groups alternately cajoling, pleading, and threatening, but the job got done. By now, I was capable of getting things done! I was also responsible for organizing and attending mechanical design reviews on each individual component. All functional groups—Project, Reliability Materials, and Manufacturing—were in attendance. These meetings could be very acrimonious. It is human nature that each group wanted to get in some comments and in the end this would and did improve the designs.

The MOL ECS requirements changed drastically at one point, requiring a new competitive proposal from Hamilton. The company generated a brand-new weight-saving idea, which consisted of using a combination **axial flow** and **centrifugal flow** fan to meet the MOL requirements. A single fan would provide the same function as separate axial and centrifugal fans. The concept was patented, and in the typical Hamilton can-do fashion, we all came in on a weekend to draw and to check a set of drawings. We had the fan built in two days by using hogged-out machined rotors, mounted on a beautifully finished wooden mount; we set it on the customer's desk, blowing air. Hamilton got the job!

President Nixon ultimately killed the MOL program, however, when he found out that satellite camera technology would produce the information needed.

CHAPTER
THIRTEEN

Taking Over the Apollo Backpack Design

I was still working on the MOL program in May 1968, when Roy Fisher, head of Space Systems Mechanical Design, called me into his office and said, "The backpack is in serious schedule trouble." His stress was visible! Then he said, "Do you think you can take it over?" Then: "For God's sake, don't say yes unless you're sure you can do it!" This apparently came from the fact that his own position was on the line.

"Yes," I said, "I can do it." I was completely confident that I could do it, provided I was given the authority needed.

Two senior supervisors had been running the backpack effort, but they kept missing design dates, and as a result, the entire Apollo program was being delayed. The two supervisors were each in their sixties and were not thrilled to be working for a thirty-three-year-old, but that was not my problem. My problem was to get things back on schedule, so we did not hold up Project Apollo!

The backpack is basically pretty simple. It inflates the space suit to 3.7 psi pressure of pure oxygen to supply the same amount of oxygen to the astronaut's lungs as on Earth. It consists of an oxygen bottle and regulator, a fan to circulate the oxygen, a lithium hydroxide canister to remove the CO_2, a water tank, a water pump, a sublimator to expel excess heat, a battery, and a radio.

As a result of testing, the Mechanical Design Group was incorporating some seventy engineering changes and updates into the backpack that would be going to the moon. Some of these were major design efforts, such as the design of a second oxygen supply, called the **oxygen purge system (OPS)**. (The greater oxygen capacity would enable the astronauts to spend more time on the lunar surface.) The OPS would have two new spherical oxygen tanks, an oxygen regulator, and controls.

Roy called me in one day and told me that NASA had asked us to calculate the area and weight of all nonmetallics in the backpack to three

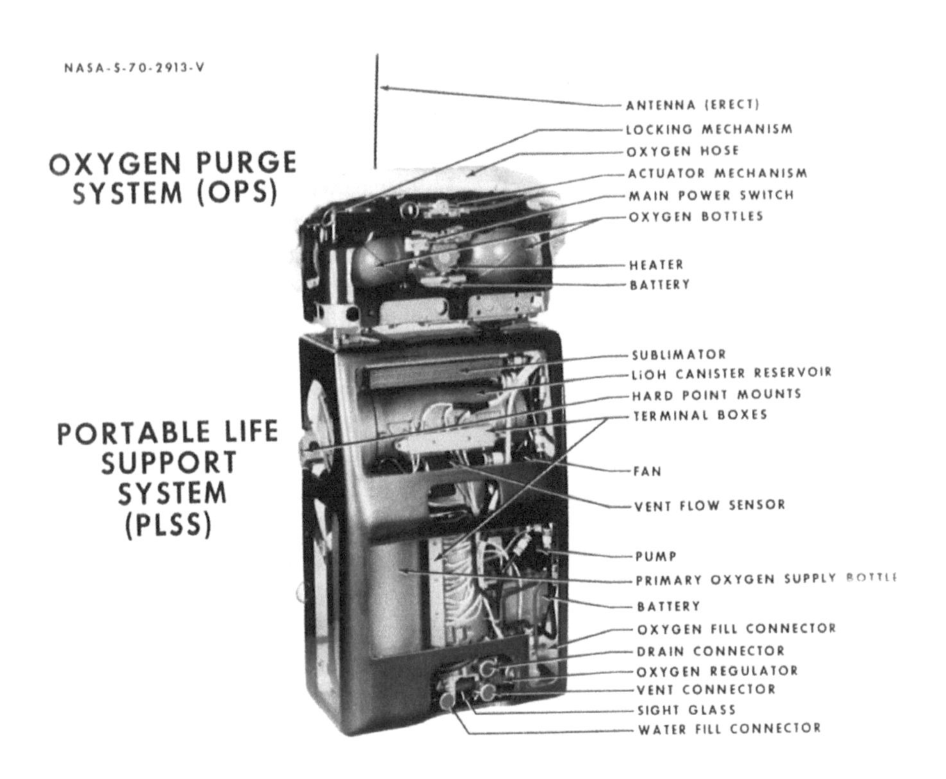

The *Apollo 9* backpack (PLSS) and OPS.
(Courtesy of NASA and Hamilton Sundstrand)

decimal places. Also, there was a $50,000 incentive (1969 dollars) to be had if we could meet a certain weight goal. I estimated it would take ten men one month to do the work. He told me to get on the phone to the temp shops and get ten men on the job by the next morning.

When I came to work the following day, the lobby seemed unusually crowded. I was pushing through when the guard said, "Mr. Smith, these men are here for you." I had completely forgotten my phone call the previous day to the temporary engineering firms. I was looking at ten expectant faces. I got them all badges and said, "Follow me." I took them upstairs and gave them to Ed Burt as their supervisor, and he put them all to work as the Weights Group.

When I took over all the schedules, I decided that, for one thing, more help was needed. I told Roy that we needed two more groups: A separate Vendor Components Group would review vendor designs for flightworthiness. We also needed a separate Soft Goods Group. Here, "soft goods" meant **thermal isolative covers**, which would be patterned and sewn up to protect the exterior of the equipment in space from radiation and micrometeorites.

I next recognized another problem: The Project Group, overseeing the entire project and setting goals, was adding design tasks without extending the schedule, thereby making it impossible for the Mechanical Design Group to meet deadlines.

Roy was an excellent, intelligent boss, and he supported me with a few minor corrections. Soon, we had five groups of ten people each, getting the work done: the Backpack Packaging Group, the OPS Design Group, the Weights Group, the Soft Goods Group, and the Vendor Components Group. I went around to each and every designer each day to check on progress. I quickly started a system of submitting design completion dates to the Project Group that were realistic and then worked with every designer and supervisor to meet them. After about a month, when we were now clicking off schedule items on time, I noticed that when I went in to see Roy, he was very relaxed and smiling and would tilt back in his desk chair during our discussions.

THE BACKPACK PACKAGING GROUP

The seventy changes that were being made to the design involved a lot of parts, lines, and tubes moving around while not allowing the package to get any bigger. One designer, Bob Waleryszak, did all this work on a drawing board with a very complicated three-view drawing. He was constantly moving everything around as last-minute changes kept coming. He did a great job, and he had unending patience. Other designers in this group worked on detailed design areas, such as new mounting structures for components that were moved.

THE OPS DESIGN GROUP

This group completed the design of the OPS, which consisted mainly of an oxygen regulator and two machined and welded **Inconel pressure tanks**, a sheet metal structure to hold everything together, and a cover. The OPS was added to the top of the backpack that was already in existence. The critical element of this design was the **tank pressure stresses** under static and fatigue conditions. We needed an overall packaging layout as well as designs for the tank support structure, a cover, and interface fittings with the backpack itself. This group was run for me by one of the two existing

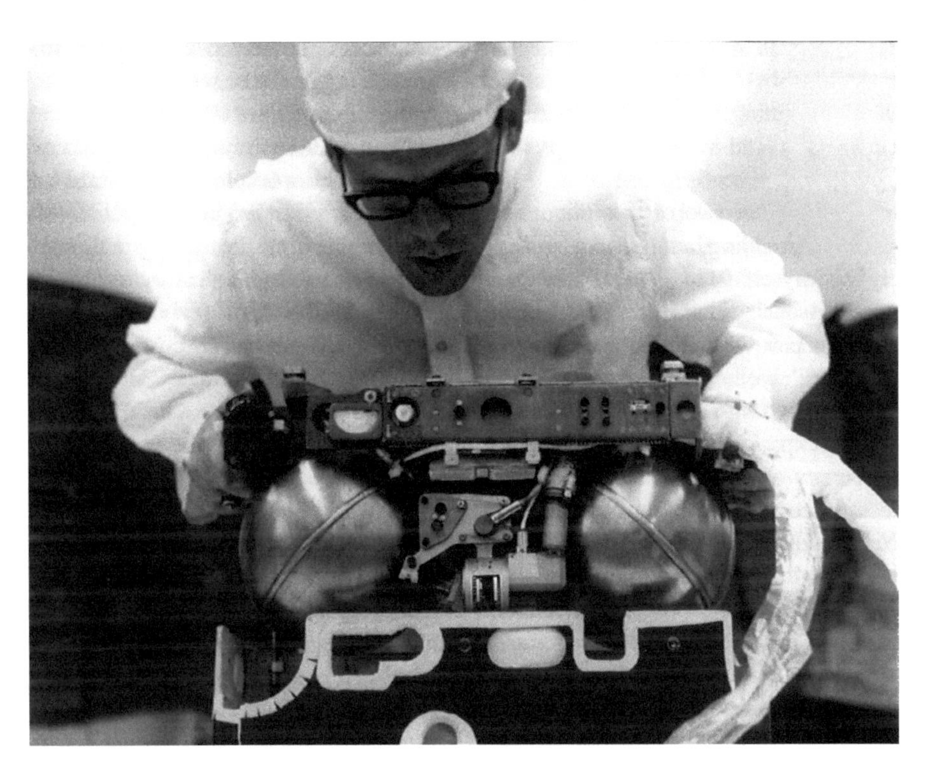

The oxygen purge system (OPS) being assembled.
(Courtesy of NASA and Hamilton Sundstrand)

supervisors. My contact in the Project Group was a very capable engineer by the name of Bill Bouchelle.

THE WEIGHTS GROUP

As I mentioned earlier, I had set up the Weights Group under a supervisor by the name of Ed Burt. Ed did a bang-up job of supervising the ten designers I had recruited for him. They did the weight calculations NASA wanted, and we succeeded in getting the $50,000 incentive award.

One of the recruited designers was an unemployed college math teacher who rode in overnight on a motorbike and pitched a tent outside of town. This particular individual was somewhat of a character in that he wore bib overalls to work while everyone else was wearing a white shirt and tie. One day, I went down to the cafeteria for lunch and found the line all the way out the door and not moving. I stuck my head in the door to see what was going on. *Oh no!* My contractor was at the cash register, fumbling through his pockets in front of a very concerned female clerk, while a whole line of hungry people behind him were waiting to eat. All he had for payment was traveler's checks, and the cashier could not take them. To get the line moving again, I paid for his lunch and asked him to pay me back sometime.

THE SOFT GOODS GROUP

The Soft Goods Group was under Art Davenport. The OPS and backpack needed to be covered with **custom insulated blankets**. As noted, the exterior of the backpack was designed to stop a .22 caliber bullet, which would be enough to withstand the impact of the largest micrometeorite expected in space. Art had been coordinating and supervising all of this work, so I just followed his schedule obligations.

THE VENDOR COMPONENTS GROUP

We exhaustively reviewed **oxygen regulators**, **transducers**, and indicators—anything that had to be purchased. I picked Lem Manchester to head up our Vendor Components Group because he had a knack for the kind of work that required digging into someone else's design. One day, we were meeting with the manufacturer of the oxygen regulator for the OPS and reviewing the design as to stresses, materials, and safety factors. Finally, after several hours of sharp questioning, the vendor said, "Look, if you don't want my regulator, I will just go home." Uh-oh. This was big trouble. We did indeed need the regulator, as our unit was designed around it and the OPS would be going to the moon soon. I did some quick backpedaling and smoothed some ruffled feathers, and we all carried on.

THE STRUCTURES GROUP

The Structures Group was under Ward Merritt. They did not report to me directly, but we worked very closely together on a daily basis. Remember, we had no computers, just slide rules. We verified stresses through **strain gauge testing**. (Strain gauges are thumbnail-size pads that can be glued to a given surface. When connected to strain instruments, they will directly measure the strain and therefore the stress in a given material.) The trick here was to have the right amount of metal in the backpack structure and the OPS tanks to be safe (sufficiently sturdy) but with none extra to run the weight up. The backpack structure was aluminum, with a known life under vibration depending on load. The structure was designed so that when it got to the moon, all the **structural life** was gone due to rocket vibration, yet there would be no danger of a failure. The goal was to minimize weight. Particularly in the case of the OPS tanks, this was serious business. In our discussions, particularly about tanks, Ward would sit back, smoking a cigar, and think and think and say nothing. Finally, he would state his conclusion, which was always sound.

CHAPTER FOURTEEN

The Race to the Moon

At some time in 1968, all supervisors were called into a meeting where the program manager, Bob Breeding, got up and said, "We want to get to the moon first." Nobody had to explain who the competition was. "Whatever you need," Bob continued, "people, money, material—you have it." In retrospect, there were reports that the Russians could have gotten to the moon first if they had had more luck with some of their equipment.

After that meeting, we were able to make some real progress. For example, when we needed two solid blocks of forged Inconel to make the prototype oxygen tanks for the OPS, I simply sent Purchasing a handwritten note with the specifications on it, flagging it as "urgent." This procedure in itself was way outside the normal routine of sign-offs and reviews.

I got a call from a purchasing agent. "Do you know what this stuff costs?" he asked. (As I recall, it was something in the order of forty thousand dollars each.)

"You know what?" I responded. "We may mess one up in machining. Get three!"

"*Whaatt!*" (I can still hear him in my head.)

"Do it."

And that was it. I had the authority. No meetings, no chain of command, just progress. Done. My verbal direction was all it took.

All the normal corporate barriers to progress were broken down, and we were going full boost on the way to the moon!

This was like war. Everything was urgent! I had to be demanding, and I was! If I was, in fact, too hard on people, I regret it, but there was no time for social chitchat. I had to lay out what was needed and when.

Bob Breeding, the backpack program manager, reported to Hamilton's vice president, Edmund Marshall. He had the total responsibility for delivering the backpack on time to Project Apollo. He had been a World War II tank driver, and his voice had a deep rumble like a tank engine. I have seen men who were real men quake in front of him. He called Roy Fisher and me into his office one day about some parts that were hung up on some matter we were involved in. "I want those f***ing parts by five o'clock," he said, "or I will have your badges!"

We were convinced and went out of his office at high speed. He got his parts, and well before five! Vice President Marshall once half-joked with someone that if they goofed up, they would get a half hour alone in a room with Bob Breeding!

Schedule pressures were enormous. In addition, in the background, was President Kennedy's promise to get a man on the moon by the end of the decade, before 1970. Someone had gone to Mexico and come back with a bull whip, which he hung on Bob Breeding's door. That said it all.

Fred Goodwin was on the backpack project for years, working his heart out. If anyone deserves to be singled out for credit, he certainly does. All of us were working sixty-six hours paid, and then we often had to come in on a Sunday. We found that after working so many hours a week for a month or longer that people started getting sick and getting behind in their personal life. I was tired like everyone else, but I always had the feeling that we were doing something special and that rest could come later. *Through it all, we never, ever lost sight of the fact that a man's life depended on this equipment.*

Walter Cronkite was the de facto representative of the media in those days. Almost every night, he would have some news about the Apollo program on TV in his unforgettably powerful and sonorous voice and delivery. I was astonished years later to hear him say that when he was reporting the manned moon mission, he did not think Apollo would work. Well, it did work! And exceedingly well!

A CLOSE CALL WITH SOME OVERPRESSURIZED TANKS

I was called at home to come in one Sunday and look at a couple of problems. The needle had fallen off the pressure gauge on a new OPS unit. There was also a fit problem with the cover, so we were banging on the cover to see why it did not fit. The next morning, we learned that the OPS tanks had been erroneously pressurized to **proof pressure**, which creates a very dangerous situation. Proof pressure is a higher-than-normal pressure applied to pressure vessels as a stress test. That was why the pressure gauge needle had fallen off. It had jumped the peg! If a tank were to fail, it would have acted like a bomb and taken out about five hundred square feet of our laboratory—and us with it!

TESTING

Hamilton tested the backpack in a regular oven—and also in a vacuum chamber with a man in a space suit on a treadmill—to prove that the backpack would work on the moon.

To collect the heat generated from walking in the space suit, Hamilton developed a water-cooled suit, which was basically a pair of long johns with clear plastic tubing sewn in so that cooling fluid could be circulated against the astronaut to remove heat. The water-cooled suit was also tested on a man in a space suit on a treadmill in a vacuum chamber to prove it removed enough heat. The Reliability Group calculated the reliability of the backpack and the LM ECS by gathering reliability information on similar elements, such as bearings, seals, etc.

MANPOWER AND SCHEDULES

One day, Kurt Barth, the head of Engineering, sat down across from me in the cafeteria. "Harvey," he said, "how many people do you have?"

"Fifty," I answered.

"What the hell are they all doing?"

I proceeded to tell him exactly what each one of them was doing, and he agreed they were all necessary.

My group completed 94 percent of all scheduled designs on time for a one-year period.

Another time, Kurt called me in for my performance review and told me, "You're doing a good job, but you're not without your failings." This startled me. Then he said, "You're too hard on people." I had been given a job that was nigh onto impossible, I did it, and now I'm too hard on people.

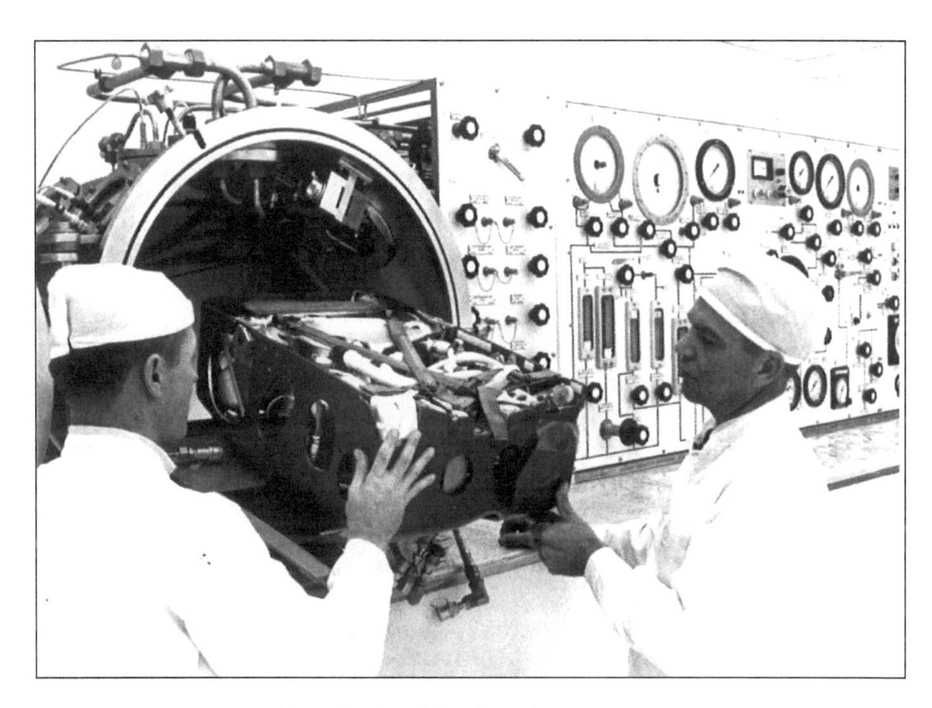

The *Apollo 9* backpack oven test.
(Courtesy of NASA and Hamilton Sundstrand)

I figured, and probably correctly, that since it was raise time, he had to find something to criticize in order to meet the raise budget. I did get a raise, but it was a skimpy one. I walked back to my desk and went right back to work without changing my ways. Years later, Kurt and I met under different circumstances, and he told me, "I always thought you did a marvelous job on that backpack."

To which I said simply, "Thank you."

EPILOGUE

The Backpack on the Moon: A Complete Success

When the backpack design changes were about all finished, I left Hamilton on good terms about two months before the moon landing. I watched the landing with my father at his house in Massachusetts. Since he is now gone, this has become a special memory.

Just a scant three months and a few days after Rusty Schweickart had used the Hamilton Standard backpack for the first time in space on *Apollo 9*, Neil Armstrong strapped on the backpack, updated for *Apollo 11* with the seventy design improvements that my group had generated with the Project Group, and clambered down a ladder to the moon.

When he got to the bottom and pronounced his now-historic line "That's one small step for [a] man, one giant leap for mankind," what my dad and I heard on our TV was unintelligible noise. Damn radio, I thought. Of all things, the backpack radio had been a constant source of problems in readying the backpack for the moon launch at Cape Canaveral, then called Cape Kennedy. In all fairness, the unintelligible noise could have been the TV network's problem that day. I would find out what he actually said a few days later.

While Armstrong was walking on the moon, my thoughts turned to the worries that I knew well. What if he falls? What if he gets sick? On the other hand, I had complete confidence in the backpack. I immediately noticed how both Armstrong and Buzz Aldrin quickly adapted to walking on the moon in one-sixth gravity. And instead of moving the higher-force space suit joints, they were sort of hopping on tippy-toes with minimum effort.

Once the astronauts were safely back on Earth, I thought, By God, we did get to the moon first, and the Hamilton backpacks worked fine! Kennedy's promise to get a man on the moon by the end of the decade had been fulfilled!

The herculean efforts over many years of more than three hundred thousand people on the Apollo program, including the hundreds of people at Hamilton Standard, paid off! It is well known, though not much talked about, that the whole worldwide political landscape would have been very different if we had not been first to reach the moon.

I am grateful and privileged to have been one of the hundreds of thousands of people who worked on Project Apollo. I have told many people jokingly that the project was so interesting, I would have worked on it for nothing, but Cecile would not have permitted it!

Ironically, the backpacks are still on the moon. They were left there to save weight on the trip back to Earth.

APPENDIX

My Salute to the Astronauts

My hat is off to all of the astronauts, men and women! They are a breed apart. Russell Schweickart has been quoted as saying, "*Apollo 9* was, more than anything else, an engineering test flight supreme." Certainly his quote is accurate, but it is also very self-effacing. To be the first man to stand out in deep space with a backpack, viewing the Earth in foot restraints and on a tether, was a monumental contribution to the space race! The astronauts look like regular people, but clearly, they are different. They know the danger, and they go off into space anyway.

One day at Hamilton, astronaut Michael Collins was in and tried on an ILC training suit. I was standing beside him (he was taller than I) as he maneuvered the shoulder and elbow joints to their limits in the pressurized suit, feeling how they worked, and looking thoughtful. At the time, I thought, He is a cool cat, just quietly doing his job with no fanfare. When the moon landing did occur, it seemed to me that he had the bigger and lonelier responsibility of staying in the *Apollo 11* command module and recovering Armstrong and Aldrin after they had walked on the lunar surface.

In 1963, astronaut Eugene Cernan was at Hamilton to monitor the LM ECS progress in a big meeting. At that point, I was a new hire, so

I just sat in the back row and listened. After a while, I could see that Cernan was asking all the right questions. This should be no wonder; I have since learned that he had degrees in both aeronautical and electrical engineering. Cernan was the last astronaut (as of this writing) to walk on the moon.

I was fortunate to meet John "Jack" Swigert in Houston on a trip with a fellow employee, Jack Kelly, who had been a fighter pilot in the Connecticut Air National Guard with Swigert. As is well known, Swigert later went on *Apollo 13*, which had problems and barely made it back to Earth. We met Swigert at his motel room, where he lived as the only single astronaut. The motel was directly across the street from the Manned Spacecraft Center. The very first thing I noticed was that Swigert was very down to earth, with no ego or braggadocio. After some initial discussion, we three decided to go to dinner. There, Swigert told us the whole fascinating story of how he had become an astronaut, a story that revealed his grit and determination.

He told us he had been working as a test pilot for North American Aviation on the **Rogallo wing**, a vehicle landing concept considered as an alternative to parachutes for the Apollo command module. A wing similar to a hang glider wing would come out of the command module and would glide to a landing. The distinguishing feature of the Rogallo wing is its triangular shape, like that of a paper airplane. Swigert related how the craft was controlled by a computer that was supposed to land it automatically. Unfortunately, the computer stalled the craft about three hundred feet up, and they crashed; both Swigert and the other test pilot were hospitalized.

Another craft was built, and this time, they were instructed to quickly assume manual control if the computer showed signs of errancy.

With the new vehicle, they did watch the computer like hawks, and—sure enough—the computer did try to stall the craft again! They quickly returned to manual control and flew the craft in for a landing.

When initial notices came out that NASA was looking for astronaut candidates, Swigert was working as a test pilot for Pratt & Whitney Aircraft, flying a modified B-17 with a single big reciprocating engine in the nose. He told his boss that he would like two weeks off to take the astronaut-qualifying tests.

"No way," his boss said. "We have work to do."

Swigert went back to his desk and got madder and madder. Finally, he went back to his boss and said, "I quit. I am going to take those tests."

"All right, all right," his boss conceded. "You do not have to quit. I will give you the time off."

He took the tests and passed everything, except the educational requirement. He did not have the required master's degree in science. He then did quit his job, went back to school, got his MS in science, and reapplied. Bear in mind that all the while, he had no guarantee that there would be any more astronaut training. When we visited him, he was ranked about halfway up in his class of forty, based on day-to-day performance.

When I came home from Houston, I told Cecile that if we had to put our personal money on someone with the best chance to overcome problems and get to the moon, it would be Jack Swigert. He was without a doubt one of the most impressive people I have ever met. He did go on *Apollo 13* and performed admirably during emergency conditions from a tank failure. After he retired from NASA, he ran for Congress in 1982 from his state of Colorado and *won*! I think some people can do anything. Consider what completely opposite endeavors space exploration and politics are. Unfortunately, he never got to take his seat in Congress, because he had become ill and passed away.

Jack Swigert was a true hero. Intelligent, courageous, and down to earth. His obituary says he was a member of the invitation-only Quiet Birdmen society—aviators and astronauts who perform dangerous and sometimes record-setting feats with little or no self-promotion. That says it all!

GLOSSARY

A&P. *See* airframe and powerplant (A&P) mechanics license.

aerobatics. The performance of in-flight stunts in an aircraft.

afterburner blast. The high-pressure, high-temperature-gas shock wave caused by afterburner combustion.

airframe. An aircraft less the engine, landing gear, and controls.

airframe and powerplant (A&P) mechanics licenses. FAA licenses that entitle the holder to repair U.S. aircraft and certify them as airworthy.

axial flow. The flow of gases parallel to the axis of the engine.

bucking bar. A piece of steel, approximately two inches by three inches by one inch, that is held on the rivet shank to flatten it when a rivet gun is applied to the rivet head.

buzz job. Purposely flying an airplane close to objects on the ground.

centrifugal flow. The flow that occurs when an impeller flings gas outward radially.

chain falls. A gearbox that is hand-driven with a chain, used to pick up heavy loads.

CO_2 removal system. In a life-support system, the use of elements that absorb carbon dioxide from exhalations.

cowling. The streamlined metal housing or removable covering for an engine.

custom insulated blanket. A protective blanket with layers of energy-absorbing and thermal-protective material.

cutoff point. The last point in the takeoff roll where a pilot can cut power and still have enough runway remaining to stop the aircraft.

dead-stick landing. Landing an aircraft with no power.

directional gyro. An aircraft instrument used for navigation.

drag. The aerodynamic force that opposes motion. *See also* parasitic drag.

dual cross-country. A cross-country training flight with an FAA-certified instructor.

ECS. *See* environmental control system (ECS).

engine pressure ratio (EPR). The ratio of the pressure at the exit of a jet engine to the pressure at the entrance of the engine.

environmental control system (ECS). A system to provide all the environmental functions needed by humans: oxygen, pressurization, thermal control, and CO_2 removal.

fatigue stress. The cyclic application of stress that, in time, will cause material failure.

fit check. Using an object with dimensions identical to a real component in order to verify that mounting features and envelope are correct.

float valve. An internal carburetor feature that regulates gasoline flow into a four-stroke engine.

flying wing. An aircraft with no tail. The inherent pitching tendency from the wing on a conventional aircraft is balanced by the tail, but on a flying wing, it is balanced by choosing wing sections that counterbalance each other.

full boost. Full throttle.

functional testing. Testing to verify that a given component or system performs as specified.

glider hitch. A hitch mounted on a glider towplane to accept the glider towline and allow it to be released from the towplane cockpit.

heat exchanger. A component used to transfer heat from one gas or fluid to another.

high-time engine. A reciprocating engine with usage hours that would indicate it is ready for overhaul.

horizontal stabilizer. A small fixed wing that is part of the tail assembly, used to control pitching.

Inconel pressure tank. A pressure tank fabricated from Inconel, a super-high-strength alloy of nickel, chromium, and iron.

inspection plate. A flat, screw-removable aluminum plate that, upon removal, enables inspection of part of an aircraft.

interface. The mating features of two components, which can include dimensional, electrical, and finish features.

iterative (cut-and-try) method. Taking what has been done before and modifying it with best ideas for improvement.

LM. *See* lunar module (LM).

lunar module (LM). Also called lunar excursion module. The vehicle designed to land on the surface of the moon.

magnaflux. *See* magnetic particle inspection.

magnetic particle inspection. A process by magnetic means for inspecting steel or iron for cracks.

magneto. A high-voltage generator that provides the ignition spark for an aircraft engine.

manned orbiting laboratory (MOL). A classified program initiated to build an orbiting space station for gathering intelligence.

micrometeorite. A minute particle that travels at high speed in space.

MOL. *See* manned orbiting laboratory (MOL).

OPS. *See* oxygen purge system (OPS).

oxygen purge system (OPS). A package designed to fit on top of the PLSS backpack and provide additional oxygen.

oxygen regulator. A mechanical device that controls oxygen flow rate and pressure to specified values.

parasitic drag. The drag created by such minor surface irregularities as rivet heads.

PLSS. *See* portable life support system (PLSS).

pneumatic computer. A computer that performs calculations using air rather than electronics.

popping it on. During wheel landings, when the wheels are about one foot off the runway, a quick forward motion of the stick until they contact the ground.

porous plate sublimator. A heat rejection device that functions in space. It works by resupplying water to ice that is frozen in small holes in a plate. As the ice sublimates, cooling is provided.

portable life support system (PLSS). A pressurization and breathing system designed to be mounted on a space suit.

potentiometer. A variable resistor used to measure electrical output as related to shaft position.

pressure suit. An inflatable garment that can be pressurized to provide the body with the same partial pressure of oxygen as on Earth.

probe survival. The continuation of functioning after an afterburner blast.

proof pressure. Exposure to higher than design pressure to prove the design before putting it in service.

random vibration. Vibration that cannot be precisely predicted, unlike sinusoidal vibration (see below).

Rogallo wing. An early Apollo capsule recovery concept wherein a set of wings in the shape of hang glider wings would be deployed from the capsule. The capsule would then be flown in for a landing.

sailplane. A glider that can rise in an upward air current.

shock strut. A cylindrical strut that absorbs landing loads.

sinusoidal vibration. Vibration that has the shape of a sine wave and varies in amplitude about an equilibrium point.

slide rule. A ruler marked with logarithmic scales, used for making calculations—especially multiplication and division.

snap release. A small mechanism mounted near the tail of a glider tow aircraft that releases a tow cable.

snap roll. An airplane maneuver in which a rapid full revolution is completed about the plane's longitudinal axis while an approximately level line of flight is maintained.

spiraling up and down. Pulling an aircraft into a tight climbing or descending turn and continuing several revolutions.

strain gauge testing. The use of glue on electrical resistors to measure strain and stress directly on the component being tested.

stress calculations. Using mathematics to calculate stresses and safety factors.

stress concentration factor. The factor for local stress increase due to change in the shape of the material.

stress node. *See* stress concentration factor.

structural life. In the case of fatigue, the predicted number of cycles that a material will withstand before it fails.

tank pressure stress. The material stress in a pressure tank due to pressure on the contents.

telemetry. A communication process by which measurements and other data are collected at remote points and transmitted to receiving equipment.

terminal dive. A dive straight down having a velocity where the drag equals the aircraft's weight plus thrust.

thermal isolative cover. A cover that protects the exterior of equipment in space from radiation and micrometeorites.

three-point landing. An airplane landing in which the two main wheels of the landing gear touch the ground simultaneously with the tail wheel.

thrustmeter. An aircraft instrument that measures and displays jet engine thrust in pounds.

torsion box. A structural design feature of an aircraft wing that sustains twisting loads.

transducer. A device that converts into an electrical signal the variations in a physical characteristic such as pressure.

trim tab. A small surface connected to the trailing edge of a larger control surface, which when moved, can move the larger surface.

weight-to-wing-area ratio. A key ratio used to design airplanes.

wheel landing. Landing on the two main wheels with the aircraft in level flight attitude.

windmilling. Engine and propeller rotation caused by air flow against the propeller.

wing angle. The angle formed between the horizon and the bottom of the wing airfoil.

wing area. The area in square feet of the wing as viewed from the top.

wingover. An aircraft maneuver where the craft is brought straight up, does a 180-degree rotation, and turns back straight down to Earth.

wing strut. A metal structural element between two wings or between a wing and the fuselage.

wing twist angle. When viewed from the end of a wing, the angle of twist that is the difference between the angle of the closest airfoil and that of the furthest airfoil.

zero-g conditions. A condition of no apparent gravity; weightlessness.

CPSIA information can be obtained
at www.ICGtesting.com
Printed in the USA
BVHW021157071119
563179BV00007B/88/P